Content-first Design

Moving Content Forward

Sarah Johnson

Content-first Design

Moving Content Forward

Credits

Case study	Shannon Geis
Cover and illustrations	Aishling Seder (except where noted in the text)
Figure 1.2	Aaron Routses
Figure B.1	© 2024 Jackson Yew. Used with his permission

Disclaimer

Trademarks

XML Press
Denver, Colorado 80230
https://xmlpress.net

First Edition
978-1-937434-86-1 (print)
978-1-937434-87-8 (ebook)

Contents

Foreword

Hello, friend,

You know, it wasn't that long ago that content folks like us felt a bit friendless at times. It wasn't unusual for content designers and User Experience (UX) writers to feel we had to spend most of our time explaining what we did, why we should be involved, and just how involving us on a project might work. We used to spend far too much of our time talking about what we did, instead of just doing the work.

These days, I'm delighted to report that things have changed. These days, we have friends everywhere. In every organization. On every team.

Our numbers are growing. And I don't just mean people with content designer or UX writer in their job title. I mean more people in all kinds of roles know that content design is essential—*essential*—to the work that they do.

I've always believed that UX writers and content designers are surrounded by friends and allies. Sure, some of them are often invisible at first. Most of them aren't exactly walking around with brightly colored name tags on their blazers that say "Hi! I'm a big fan of content design." Most of them aren't even wearing blazers. Which is a bit of a shame. I love a sharp-dressed colleague, don't you?

Right. Friends. We've got loads of 'em. And they need our help.

Because even our most enthusiastic partners can struggle sometimes to define what we do. And to explain just how they and their teams should plan to work alongside us, and how they can get the most out of us when we're involved.

Now that we're at this point in the journey, it's time to give them that plan. That map through the woods of content design.

First, let's take a moment to enjoy how far we've come.

When I started out as a new UX writer, I literally didn't know a single other person who did what I did. I'd tell my friends I was a UX writer. Not a lot of people knew what those words even meant. Least of all the designers, engineers, and product managers I was trying to work with. How exactly was I going to help them ship better products, help more users, and connect the dots across a complex product landscape, if all I did was work on "the words"?

Well, I love a challenge. And I bet you do, too.

UX writers and content designers started getting to work. We moved—sometimes slowly, sometimes quickly—from late-stage proofreaders and copy editors of strings into real active partners of our peers in UX design. We started doing more and more foundational work. Content audits, user journeys, empathy mapping, product narratives, prototyping, content-first wireframing, and more, always keeping *content* first.

That was a watershed moment, right there.

When we finally realized that the content can come first in a design process, we saw that we could be starting with the real building blocks of the solution—the ideas, concepts, metaphors, mental models, and goals—not just the boxes and bubbles that would later contain them.

In retrospect, it's not even a little surprising. When you think about it clearly, putting the content design first in a project makes perfect sense.

But it's still a new practice for most teams I know. That's why this book is landing at just the right time.

More people than ever are realizing that the content *is* the experience. More teams than ever are starting to ask for—and get—the chance to work with a content designer by their side. At the same time, more content-ready product designers are trying to learn how to infuse more of the practice of content design into their work. And they're not alone.

All those friends and allies we talked about a minute ago? They've always been eager to adopt the lens of content design in their work. You don't need to wear the title of content designer to see the value in our practice, or to want to do what we do.

I don't care if you're a product manager at a tiny startup or the Director of UX at a multi-billion dollar unicorn. I see you. You're holding this book, too. And you want to know if content-first design can help you create better products, too.

The answer is yes. So let's skip right to that part.

Content-First Design gives you a practical approach to solving the challenges we all face today. How to align with internal stakeholders. How to set realistic yet ambitious goals. How to plan, prototype, test, and iterate in a content-first way.

Whether you're a solo content designer supporting multiple products, a product manager trying to embed content thinking into your process, or part of a mature content team looking to scale your impact, this book offers clear, actionable guidance that will get you there.

And yes, it's more than just inviting certain charming people to an earlier meeting. Content-first design is a blueprint for action that *anybody* can use.

Think of this book as your map through the complex terrain of modern product development. You'll find regular rest stops for reflection, scenic overlooks for perspective, and plenty of shade for rest and contemplation.

The path is well-marked, but the journey is yours.

But every journey is better when you set out with a friend.

You know what they say: If you want to go fast, go alone. If you want to go far, go together. So pack your bag, lace up your boots, and bring along your whole team. The sky is clear, the path beckons, and so many fellow travelers are waiting to walk with you, just around the bend.

Welcome to the journey. Happy travels, friend.

Beth Dunn
Author of *Cultivating Content Design* (Dunn 2021)

Preface

Content-first design is the process by which research, exploration, and evaluation of content requirements and user needs inform the structure, layout, flow, and visuals for a digital product.

A content-first design approach is not a new concept. While the focus of this book is a content-first approach to web and mobile design, "form follows function" is a tried and true formula we see in countless other industries. Just as architects consider their clients' visions and accommodate construction and budget constraints when designing spaces, understanding the user, the user problem, and the user's goals is paramount in developing successful products.

While I demonstrate a methodology that has worked for me, I want to point out that this is a flexible concept and can be adapted to your organization in many ways. Also, this process can be introduced in phases, even tiny phases if that works best for you. That said, read on with an open and critical mind, see what you can learn, and I hope this book helps you improve your design process and ultimately your customer experience.

I first began to think about content-first design when I was working on a re-brand of the Teachers Insurance and Annuity Association (TIAA) website, mobile experience, and app. In preparation for the user experience design, we conducted countless customer interviews, listened to call-center conversations, held empathy mapping exercises, and organized regular usability testing sessions.

From this pre-design work, we learned about our customers' fears, aspirations, and goals for conducting business online, as well as how they navigate and speak about their finances. By sitting through the research phase, we were able to come up with a list of principles to help guide us in writing about financial services online.

As I dug in to work with designers, I realized that we were creating a conversation with customers, many of whom are retired seniors, guiding them through potentially complex financial tasks and simplifying complex ideas into easy-to-follow instructions, reassuring them along the way, as we now understood their fears. Together with our user-interface designers, we used this pre-design work to create a content-first design for our website, mobile experience, and app that created a conversational interface with our customers.

The outcomes were off the charts. The percentage of customers successfully completing tasks went up by as much as 76.4%. This told me that looking into content-first and gaining a deep understanding of the user helped us deliver successful solutions.

Content-first design happens in collaboration with the UX designers, UI designers, accessibility experts, product managers, compliance or legal experts, and other key stakeholders who can help further your exploration into requirements and user needs to create best-in-class product solutions.

One of the things we learned at TIAA was that users require simple language to explain complex ideas, no jargon, no figures of speech, nothing that can be misunderstood across geographic, socio-economic, sexual orientation, race, and other factors weighing into a user's background. Today this is called *plain language* and is a guiding principle in UX writing, but this work at TIAA took place prior to that thinking and the development of plainlanguage.gov. It also refers to inclusive design, which means that anyone who visits the site or app understands the message, next best actions, and how to achieve their goals.

Why content-first design?

One of the biggest frustrations of content designers is that we are brought to the table too late in the process, left out of strategy meetings, and asked to fill in the blanks for a finished design. We're thought of as glorified copywriters. If I hear content referred to as "copy" one more time, I'll go bananas.

Frustration continues when there is no collaboration between content, design, and other stakeholders. During a content-first design process, stakeholders participate in content discovery and definition. Additionally, stakeholder participation emphasizes the complexity behind content-first design and educates them about the measurable results you can achieve with this process. This both helps you build consensus and allows everyone to gain deeper insight into the user, the business, and the product you're working to develop. Educating stakeholders is critical to gaining trust so that you can get support earlier in the project design process and get a seat at the table as early in the project cycle as possible.

So where did the idea for this book begin? In 2018, I was asked to create and teach a 2-day course for the Bentley University User Experience Design Center. As I considered what would be most valuable to future content designers, I looked at projects I'd worked on and found that defining content first led not only to the highest performing content but also to the most successful products.

The goal for this book is to provide a framework for content-first design that you can use to create more meaningful user experiences and, hopefully, get you into the project-planning process earlier. When you demonstrate, both qualitatively and quantitatively, the value of this process, you'll be better positioned to advocate for content-first design across the organization.

In *Content-first Design,* I encourage readers to read, listen, and learn to think critically about the ideas presented here. As the field of user experience (UX) content evolves, one thing you'll learn is that there is no one right answer—just the best answer for you and your situation. This book is here to help you find *your answers* and to invite you to participate in the evolution of content-first design.

In this book, you'll hear from content experts at Expedia Group, Google, Microsoft, Meta, and other industry leading companies. You'll benefit from a wide range of opinions, thoughts, and experiences. It's up to you, reader, to decide what you think. I hope that by laying out this methodology, and these wide ranging thoughts and experiences from experts, you'll learn the value of putting content first.

Who should read this book

Whether you're new to content design or already working in the field, this book will help you push the boundaries of your knowledge, creativity, and ability to collaborate with other team members including UX, product designers, and business stakeholders.

Content professionals working in the field

Even the most seasoned content practitioners are always looking for ways to learn more about content-first design. While *content-first design* is a phrase that's tossed about a lot, it hasn't been written about in depth or with suggestions of any clear process. One of the best ways to pick up new skills is to look at your work from a new perspective. If you're a content design nerd, you'll find some ideas and tools here that can help you move forward in how you think and work with content-first design.

New content professionals

Okay, so you have a lot to learn. *Content-First Design* will put you ahead of the game with a ton of information and examples that you can learn from. This book will take you through every phase of a content-first design project and give you the tools you need to be successful.

Learn what to expect from other members of your team (both in and outside of user experience), process and project flows, and how to measure success.

Content-curious professionals

There's a wealth of information in this book for designers, researchers, accessibility experts, product managers, and others charged with creating or designing digital products. Here you'll find the keys to start shaping experiences and conversations around the content your users need, based on your business requirements and solid content design.

What this book is not

While I include a quick tutorial on UX writing, this book will not teach you the ins and outs of the UX writing craft. Rather, it will teach you the process whereby you arrive at the best possible content design for the problem you're tasked to solve for your users. I provide book recommendations here and there to help you gain deeper learning around UX writing. In fact, you'll hear from some of the authors in the "From the experts" sections.

How this book can teach you to learn

One of the most exciting things about the field of content is that it's rapidly developing and changing. Even seasoned UX content practitioners are always reading the latest books, listening to podcasts, and searching the web in order to improve their own knowledge and work.

This book aims to curate a diverse group of perspectives about working in content, introduce you to some new ways of working, and get you into the rhythm of continually learning and thinking for yourselves as UX content professionals; this is key to your success and the success of your projects. You'll see some divergent points of view in this book, because different jobs require different ways of thinking and working. By exposing you to a range of ideas and perspectives, I hope to help you form your own base of knowledge and opinion about what content-first design means for you and your organization.

If you read this book with an open mind and a critical eye you will be able to sort out your thoughts and strategies. What works for you? What would you add or disagree with? It's all fair game.

How this book is organized

The largest part of this book describes how I use the four phases of the double-diamond design methodology in the content-first design process. Those four phases are: discover, define, develop, and deliver.[1]

To further illustrate the process, I've included a case study, developed by Shannon Geis, that follows a hypothetical bank, Tidal Bank, as it uses the content-first design process to develop a new website. You can find resources to go along with the case study at contentfirstdesign.com/resources. These include a PDF of the case study and the Figma file for the design, so you can follow along as you read.

You will also find sidebars that provide the perspectives of experts in the field, extra details, and explanations of new or less-well-known concepts.

 From the experts sidebars let you hear directly from experts in the field who weigh in on the ideas presented in the book and share their direct experience.

 Dive deeper sidebars are bits of information to help you gain a deeper understanding of concepts and tasks covered in the book.

 Concept corner sidebars explain ideas or language that you may not be familiar with, but that will help you understand more clearly what you are reading.

---

[1] There are other design and development methodologies that work well with content-first design. I discuss some of those other methodologies in the book, but I focus on the double-diamond process.

Stay connected

As you read through this book and think about how content-first design or the methods and principles described here could help your team, please visit contentfirstdesign.com for news and information. Send your comments to sarah@contentfirstdesign.com; I want to hear from you.

Acknowledgments

Thank you first to my editor extraordinaire, Richard Hamilton, and XML Press for their continued support throughout the process of writing this book. Also to Kevin Nichols and Scott Abel for their ongoing support. Thank you Gregg Almquist, who secured my position teaching content strategy in the Bentley University User Experience Design Center through an introduction to Bill Gribbons, and thank you Bill.

This book would not have seen the light of day without the early support and encouragement of Melinda Belcher, Executive Director of Experience Design at JP Morgan. Many thanks to Aishling Seder for illustrations and editorial feedback and to Shannon Geis who wrote a compelling case study and provided editorial support.

I'd especially like to thank all of the experts who contributed their experience in order to make this book one of shared voices creating community around content-first design. Thank you Beth Dunn for taking the time to write such a thoughtful foreword. If I've left anyone out, please forgive me. A book is not made in isolation. It takes a village, and I'm very grateful to my village.

Sarah Johnson
February 2025

CHAPTER 1

Content-first design—because content is design

Any digital experience is about creating a conversation with users. Just as you can't have a movie without a script, it would be tough to have a conversation without words—or the images that represent words, such as photos or videos. How can you create a digital conversation with users if you have no idea what you need to say to them, what they need to hear from you, what they're looking for, or what they need to know? Of course great content is about more than what you want to say to users. Great content provides a frictionless, informative experience that helps users complete tasks as easily and intuitively as possible. Putting content first helps create frictionless experiences that provide user satisfaction and improved metrics toward reaching business goals.

Friction: In experience design, friction is anything that prevents users from accomplishing their goals or carrying out tasks. Friction can be as simple as an overlay covering a button, unintuitive wording on a product page that throws users off, or worse yet, a list of optional questions in a checkout flow.

Content friction includes confusing content, content above a specified reading level, inaccessible content, poorly written links, lack of information that would allow users to make informed decisions, and missing information. These are a few examples of content friction. In Chapter 11, *Test and measure content at every phase* (p. 109), I talk about testing content, which is how you'll find those unintuitive places that stop users in their tracks.

Many times, I've been brought into a product design process and asked to plunk some content into wireframes filled with *lorem ipsum*. Once I saw the wireframes and thought about the conversation we were trying to have with the user, we often had to go back and rethink, revise, and redesign. If the team had brought content designers into the process earlier, understood the business requirements, and had time to conduct strategic work, we could have created a more user-centered design and delivered a better, more efficient, and less expensive solution.

Every time someone says "oh, we don't need words yet" I have to resist punting them out of a window.

The words are an expression of the solution, the last 20%, but we also need to do the 80% that comes before to know wtf to write. There is no magic wand that makes the words all better.

—Relly Annett-Baker, Head of UX content design for Google Corporate Engineering, from a post on linkedin.com

A content-first approach can save time, money, and resources on any product design project. It can be implemented all at once, or in phases to smooth the transition in a new process.

This begs the question, *why isn't a representative from the content team included at the beginning of projects?* Your guess is as good as mine, but here's mine: as user experience evolved from the early days of the internet, designs were initially considered information architecture, creating hierarchies and structure. As we moved forward, the concept of a *user experience* required that we look beyond the layout and structure of information toward how we want to drive the user through completing a task. Notice how these are non-content related steps in an ever-evolving, expanding, and fascinating field. Content was typically the last step to design work; the frosting on the proverbial cake.

Now that we've learned more about what we are trying to do for users and how to help them, it's clear that content is primary to a successful user experience. What then is the resistance to getting content involved early? It is most likely a basic lack of understanding about what a content designer brings to the table.

Since the beginning of time, User Experience (UX) teams, including content and Content Management System (CMS) developers, have believed in the misconception that content is simply words, with no concept as to the thinking, analysis, research, and strategy that goes into those words and how those strategic words can lead users through an experience more effectively than simply relying on tasks or actions the user must take. In Chapter 3, *The discover phase* (p. 25), and Chapter 12, *Stakeholder buy-in* (p. 123), I look at tools and strategies for educating your stakeholders on the value of getting the content team involved early.

As we were just saying at our stand-up meeting the other day, "Lorem ipsum dolor sit amet, ius ne dicit oblique neglegentur, right?" Allow us to translate: "You shouldn't just wait till the last minute to put the real copy into a website, app, or some other digital product, right?"

Just as the concept of mobile-first has helped designers prioritize their work, a content-first approach does the same thing for the entire team. When you start from the beginning with actual content, prototypes and user tests are more accurate, and your chances for success go way up. It just makes sense, doesn't it?

—Larry Asher, SVC Seattle, from a conversation with the author

Now let's go back to that conversation you're trying to have with your user. You might already know what you want to say: "Buy this widget!" or "Sign up here!" But how do you know what your users need in order to buy your widget or sign up for your service? Designing a streamlined, thoughtful conversation that guides people on a seamless selection journey requires content research and analysis before UX design gets started. UX design will collaborate and participate in content research and analysis, but these efforts will be led by the content designer.

Words are essential for helping users accomplish their tasks, and by thinking about them while you sketch, you'll uncover problems early and be able to move faster later.

—Biz Sanford, Shopify[1]

Putting content-first means evaluating and gathering information about content before digging into design. It means implementing a multi-phase process to create an understanding of your users, how they think and speak, what they want, and how they expect to get it. It means looking at business goals and how to achieve them by putting the user first.

---

[1] "Words and the design process" (Sanford 2017)

The idea of content-first design was introduced by Jeff Zeldman, the founder of *A List Apart*. He wrote that in order to create the right layout and structure for any digital experience, you have to know what the content is first.

zeldman ✔ @zeldman · May 5, 2008   ⋯
Content precedes design. Design in the absence of content is not design, it's decoration.

💬 81   ↻ 2.3K   ♡ 2.5K   ↑

I think of content-first design as meaning-first design. What something means is what makes it valuable. It doesn't matter how formally interesting or elegant a design is if it is meaningless or if some aspect of the design—the interface, the style, the behavior—impedes understanding. So, design should begin with an understanding of what you want your product, service, or system to mean to the intended audiences or users. From that you can determine what content conveys that meaning, and its attributes and requirements. Only then should you think about the design that best supports that meaning and that type of content.

—Erika Hall, Author of *Just Enough Research* (Hall 2024)

Another reason for content-first design

When you design without putting content-first, the digital experience you're creating becomes like that movie without a script that I mentioned earlier—chaotic, stressful, and expensive. Content shoved into an experience with no concern for the user's journey and needs, is usually:

- Not useful or usable
- Not targeted to the users' needs, voice, or intent
- Difficult to navigate
- Redundant
- Inconsistent or incomplete
- Irrelevant, ineffectual, or out of date

The result is a design that is not user-centered because the designers weren't cognizant of user needs as they tried to create a conversation with the user. The user experience itself is often flawed, requiring extra content and extra clicks, leading to higher drop-off rates. You have seconds to capture someone's attention in the digital space. If your content isn't well-researched, tested, and presented, users will get confused, bored, or frustrated and move onto something else. They won't trust you, and you will have lost their attention. This reflects poorly on your brand.

Without a user-centered content design that analyzes existing and potential content and without pre-discovery work with an eye toward content, meeting business goals goes right out the window. If you explore user needs around content and move forward with a clear vision, content-first design will help you deliver experiences that users connect with and understand right away.

The right content and design help you build a product that creates meaningful relationships between your users and your brand.

 Content supports a customer's needs along the entire journey, from discovery to loyalty, and your business will suffer if the content is failing your customers.

—Paula Ladenburg Land, Principal Consultant, Enterprise Knowledge, LLC, author of *Content Audits and Inventories* (Land 2023)

What makes for good content-first design?

- Follows a detailed process outlined at the start of the project
- Elevates the impact of content so that each task is easier to carry out for the user
- Places the user at the center of the digital product design process
- Evaluates content requirements and digs into the mindset, emotions, and capacities of the user to create digital products that are intuitive, effective, accessible, inclusive, and successful for all users and the business
- Provides prototypes with real content for stakeholder review
- Improves collaboration with UX, product, accessibility, and other stakeholders

What happens when there is no content-first design?

When there is no content design, the digital experience will be a mess. Instead of guiding the user, content will potentially be:

- Difficult to find
- Structurally confusing
- Inconsistent or incomplete
- Out of date
- Irrelevant

Users are likely to either get lost and frustrated or abandon the experience entirely, decreasing confidence in the company and reducing sales, donations, and information gathering.

 Everything—UX, design, development—flows from content. If we don't know the meaning of the experience we're creating, we cannot design that experience. Content speaks most directly to intent and meaning in a digital experience. If you try to introduce content after, you'll find all the leaks in the boat and have to, likely, do some very expensive redesigning.

—David Dylan Thomas, Founder and CEO at David Dylan Thomas, LLC, author of *Design for Cognitive Bias* (Thomas 2020)

Figure 1.1 shows the home page of gatesnfences.com and demonstrates what can happen when there is no consideration for content-first design.

- The structure is confusing
- Two menus that don't guide the user
- Large blocks of difficult to read text
- And simply too much to absorb
- No clear path for the user and no next-best-action
- A level of disorganization that makes it difficult if not impossible for the user to navigate.

What else do you see in Figure 1.1 that you can call out as lacking content design?

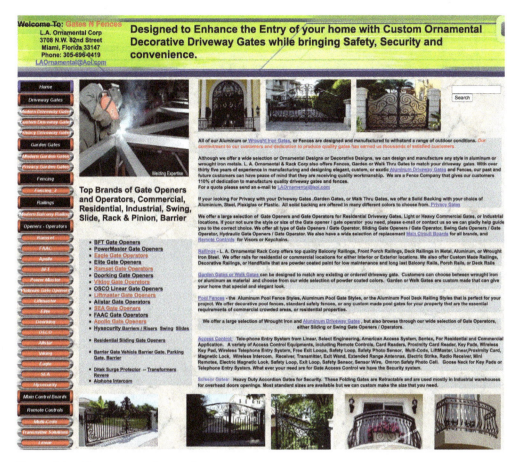

Figure 1.1 – Home page of gatesnfences.com

Once you learn the product building process, you start to notice when the apps, websites, and other digital products you're using didn't consider content early. It's when content is clearly being used as a band-aid to fix a bad design or engineering decision.

One time I was brought into a project three days before it was scheduled to ship some upgrades, and the product manager said "Andy, we're changing this key term from X to Y, because we've acquired this other company and we're incorporating some of their functionality into our product. They already have a feature called X, and they have more users, so we're changing ours to Y."

(And yes, I am being intentionally vague here.)

So what could I do? If they brought me in as early as possible, perhaps we could have done a system audit, found instances of the term X, and worked on some user education moments to teach users that they were going to be Y. Or maybe we could have worked with the product team on the other product and figured out a way to use the term strategically or found an alternative.

But that didn't happen. Instead, I wrote an alert dialog that popped up when the update was applied that said something like "We're now calling X by a new name—Y!"

It was a bad user experience that likely frustrated and confused a lot of users. And it was all because we didn't think about the language we were using early enough in the process.

—Andy Welfle, Principal Content Designer, Microsoft, and co-author of *Writing is Designing* (Metts & Welfle 2020)

What could happen instead

Figure 1.2 is an example by Aaron Routses from the Content Design course at Bentley University. For his final project he reworked the page from the gatesnfences.com website shown in Figure 1.1.

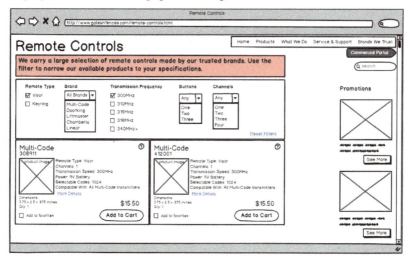

Figure 1.2 – Re-design of Figure 1.1 by Aaron Routses

Aaron worked on this design as part of his final project. Even without the benefit and expertise of working alongside a UX designer, he was able to create an experience based on content alone. While this is a rough draft that will be worked on with stakeholders, tested, and put through the mill, it provides a starting point for creating the conversation you want to have with your users.

You can see that he structured the content, reduced the amount of text on the page, and used language to lead the user through an experience. Rather than tossing too much information at the customer, he allows headers, sub-headers, and brief text to help the user digest the information and guide them through an experience of shopping and buying products. This example illustrates how content-first design can help you deliver best-in-class digital products.

The next section ("Case Study: Introduction") introduces a case study created by Shannon Geis for this book. The case study follows a fictional bank, Tidal Bank, as they update their customer website using content-first design practices. Most chapters contain an installment of the case study that illustrates the concepts introduced in that chapter.

TIDAL BANK

Case Study: Introduction

Tidal Bank provides a variety of banking and savings accounts to customers. They want to create a new onboarding flow for their mobile app that explains the different savings products they offer. The bank wants to educate their customer base, which is primarily adults in their 20s who are just starting to think about their financial needs. Research the bank has conducted shows that these customers aren't always sure what products are best suited for them and want more education and guidance before making big financial choices.

The bank wants to give customers a better understanding of the financial products they offer and provide guidance on how to choose one product over another. However, the team is unsure of how to organize this information, what order the information should be in, and how much information a user might need to make informed decisions.

They also want to use the process of customers setting up a login to help them choose the right product(s) for their financial goals and then start opening an account(s). The team isn't sure how to move from the product information to a sign-up flow. They are also not sure how much information they need from users to get started with the account opening process.

Tidal Bank's current website looks like Figure 1.3. The following chapters will show how the website evolves using a content-first approach.

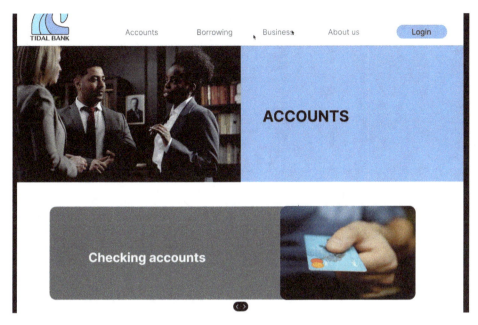

Figure 1.3 – Tidal Bank current website

CHAPTER 2

Your team's content-first design process

What is a content-first design process?

I first began to think about developing a content design process when I was at TIAA and we were re-branding the site, app, and mobile experience. The need for a structured methodology became apparent when I realized that stakeholders such as product owners, business, compliance, research, and designers didn't know what to expect when working with the content team.

At that time, the content team was typically called in once wireframes were in place and approved. We based our content process on a typical project experience. The process was to write to the wireframes based on user research and work with business and product owners on approvals, before we went to compliance.

It wasn't until I began to develop a content design course for the Bentley University User Experience Design Center, that I took the time to reflect on what I wanted to get across. What worked best in content design?

I thought about the research conducted before the re-brand and how it influenced content. I knew that we were frustrated because the team was brought in late to projects and not included in the initial design phases. It was like writing in limbo.

What if the content design team conducted research before a project kicked off? Isn't digital design about a conversation with the user? What if that conversation led the design process? It only made sense. To me, at least.

I constructed the course by having students start with research methods I learned at TIAA, such as empathy mapping, conversation mapping, listening to calls at the support center, and more, which you'll learn about in this book. Then after teaching them some UX writing basics, I have them redesign a site using content based on what they've learned.

And it worked!

The students developed great websites that with a little tweaking would be very successful. We all learned a lot about putting content first. I realized that this was a process for UX design, and I developed the process you'll read about here.

The content-first design process has the following elements:

- Defines what tasks the content designer does at each stage of a project
- Pinpoints where the content designer or team interacts with project stakeholders such as UX design and product, i.e. meetings, workshops, reviews
- Identifies what the deadlines and deliverables will be at each point in the project process

An effective content-first design process will help align team members and provide context for how best to collaborate with the content team at each stage of the project.

Some projects won't use all the steps of a content-design process. And sometimes a project might be larger in scope and require a more refined and detailed process. Either way, having a common foundation to build from can make the difference between a cluster of miscommunication and a successful project outcome.

What happens when there is no defined content-design process?

- There is nothing in place to guide content designers and teammates forward
- The team has no way of understanding how to work effectively together
- Content gets brought to the table too late and with too little information
- Stakeholders don't know what to expect when working with content, and they don't know what's expected of them at various touchpoints in the design process
- Misunderstandings and missed deadlines can derail a project
- The content practitioner and colleagues are left feeling out of sync with the team or simply overwhelmed by a barrage of requests with no clear way of meeting those requests

Benefits of a defined content-design process

- Clearly communicates expectations on all sides of the table
- Demonstrates the value of the content team by including stakeholders
- Helps estimate scope and timelines, thereby limiting scope creep

Concepts behind content-first design

There are as many ways to design a content-first process as there are organizations creating content. The trick is to identify and delineate the process that will work best for your organization to create best-in-class products. Let's look at a few ideas before we dig into a real-life content-first design process.

- Design thinking
- Human-centered design
- Double diamond

While I use the double-diamond approach in this book, here is background on other approaches so you can choose what works best for you and your organization.

Design thinking

Design thinking has deep roots in a conversation that went on between designers around the world, it and has now established a footing in content and UX design. Design thinking is an iterative approach to designing that puts the user center stage. It uses creative brainstorming based on learnings about the user and testing and retesting with actual users. This process uses quantitative and qualitative data from an empathize phase, then lots of brainstorming, followed by testing.

This iterative process repeats until a successful solution is found—one that reduces friction and gets rid of pain points for the user.

The design-thinking philosophy involves the following:

- **Solutions-based:** rather than focusing on the problem, design thinking focuses on finding solutions
- **User-centered:** design thinking puts the user at the center of the experience
- **Hands-on:** every point in the process will get you working to conduct research with users, generate ideas, and test prototypes
- **Iterative:** repeats steps of the process or even the whole process until you find the right solution. Finding the right solution is the name of the game.

Figure 2.1 walks you through what happens at each stage of the design-thinking process.

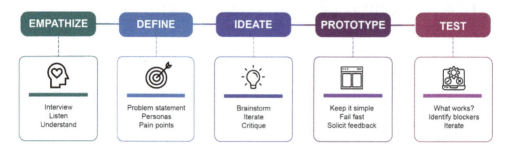

Figure 2.1 – Design-thinking process

Empathize: In order to understand your end users, conduct research to uncover their issues, preferences, likes, dislikes, thoughts, feelings, and more as they click through a digital process. You do this with discovery work. In this phase of a project, you ask questions about the users and their experience. You work to uncover and avoid your own assumptions about users and potential solutions, as this can potentially lead down a quagmire of more problems to solve later. In some organizations, users are invited into the process to play with prototypes, brainstorm, and generate ideas. This is fun for everyone and super useful in uncovering unexpected pain points, points of delight, and points of friction in the flow.

Define: You can define the project by focusing on what problem the team needs to untangle for the user. Beware of trying to name a solution before naming the problem. When you get ahead of the problem and avoid the rest of the process, you can create costly mistakes that the process could've helped you avoid. A problem statement acts as a north star for designers, product owners, and other stakeholders as they explore a range of solutions that get at all aspects of the problem.

Ideate: Once you understand the user and have a grasp on the problem, it's time to get your creative hats on, design as many resolutions as you and the team can think of, and put them together for your intended user. One way to do this is in a workshop where teams get together and put ideas on sticky notes on a board, either virtually or in person. Some teams include users in the process. There's nothing like hearing straight from the source. When you understand how users experience the problem head on, you gain valuable insight into how to approach finding a solution. You'll find that the users' suggestions can prove most valuable of all.

Remember, this is the time to put assumptions and judgments aside. You want to get everyone's ideas on the board and be sure not to shut any possibility down. Brainstorming is creative time,

time to let her rip, so enjoy the creative process, and once you've gotten everything down and created a possible solution, you move onto the prototype stage.

Prototype: Prototypes are low-fi design experiments created to try out possible solutions, or parts of solutions with actual users in the next step. Best practices suggest that having a variety of prototypes to address different problems can help you gain the most insight from the test and iterate phase. These can be word-based or include wireframes with rough or first draft content.

Test and iterate: This phase will help you understand where the friction and pain points are and enable you to make improvements as you go. Users who deal with these pain points test the prototype over and again to ensure that it solves all facets of the problem you're trying to solve for.

The goal of this phase is to learn about gaps and pain points, not defend your solutions. Be curious. Be open, listen, empathize, and understand. Take this opportunity to gain insight into the user's experience. What works for them, what doesn't work, and why is that?

Benefits of design thinking

With a deeper understanding about the user's experience, you'll be able to design content that works, rather than content you tried to push through because you thought it was right. This improves the overall user experience, removing friction and pain points that can lead to frustration and abandonment.

Design thinking is a proven method for addressing problems in digital products, including content, that helps you move quickly while keeping your solutions user focused. Consider design thinking for any content and design problems you're facing and you'll be surprised by the results. I can assure you of that!

Human-centered design (HCD)

Human-centered design (HCD) is an approach to creative problem-solving in technical and business fields. The origin of human-centered design is often traced to John E. Arnold, who founded the Stanford University design program in 1958. Professor Arnold first proposed the idea that engineering design should be human-centered.[1]

As we know it today, HCD is a process that focuses on developing a deep empathy for the users you're trying to reach. It takes into consideration your users and their needs, requirements, and

---

[1] "What is human-centered design? Everything you need to know" (Vinney 2023)

context. In human-centered design you seek to understand and solve the right problem for the user. You work to get at the root of that problem so that the symptoms of the problem don't return.

Human-centered design is about cultivating deep empathy with the people you're designing with; generating ideas; building a bunch of prototypes; sharing what you've made together; and eventually, putting your innovative new solution out in the world.

—IDEO Design Thinking[2]

In order to explore the interconnected parts of any system, as well as gain insight into the user's experience, you do small and simple interventions. This means iterative work: prototype, test, refine. Through rounds of this process, you hone in on small solutions that add up to uncovering problems and delivering the right solution at the right time and the right place. HCD is iterative, measurable, and repeatable.

The goal of HCD is to create a design derived from an understanding of users, their tasks, and their contexts. For example, imagine a 77-year-old user who wants to check her retirement balances online, then potentially make a withdrawal. She's not comfortable with technology, so her son set up the app for her on her phone. What problems might she run into? How can you begin to solve some or all of those problems?

In order to keep designs on track with user needs, users remain part of the process at each phase. They teach you about their context of use and you learn their requirements. The users participate in the design process and help evaluate designs. With the user at the center of the process, you can move quickly, efficiently gathering information, refining, testing, and delivering.

With guidance from users and frequent testing you continue to move toward your final content design. Some of the ways to evaluate your designs through each iteration include: usability testing, user interviews, and use cases that outline every possible trajectory a user can take and how the product will respond. I cover these research methods later in the book.

Figure 2.2 shows the three phases of the human-centered design process.

---

[2] "What's the difference between human-centered design and design thinking?" (IDEO)

Figure 2.2 – Human-centered design process

Inspire: First you learn straight from the users by immersing yourself in their experiences and their needs. Your goal is to develop a deep empathy with the user. You explore their experiences, goals, and preferences. You learn about their knowledge levels, what makes them feel comfortable, and what does not. Once you are able to empathize with the user, you can begin to understand their problems and some things you need to consider in addressing the problem. You explore the users and their needs through research, conversations, interviews, usability testing, and other research activities that I explore later in the book.

Ideate: This is the most generative phase of HCD, during which the team designs solutions to see which most effectively answer the needs of the user. During this collaborative process, team members name every idea and then evaluate each to create a potential solution. The potential solution is generated through a deep understanding of the user gained from the inspire phase. The solution gets prototyped with a low-fidelity wireframe[3] and first-draft content so that the user can take a test drive and provide insight into further iterations of the design. The idea is to learn from mistakes and make improvements to develop better solutions.

---

[3] A low-fidelity wireframe is an early, rough sketch of a website.

Implement: Now it's time to put that prototype to the test and build a working demo that you can test and further iterate on. Then the product gets built by development and evaluated in a multitude of ways, including qualitative and quantitative metrics, usability studies, and user interviews. This evaluation once again helps designers revisit and refine their solutions, and the process continues again until the right solution is developed for your user.

Principles of HCD
There are four principles to HCD, with the first principle perhaps the most important:

- Get at the core problem you're trying to solve.
- Be user-centered: empathize with users through deep research into their context, problems, and pain points.
- Use an activity-centered systems approach: activity-centered looks at the activity of the user, in context, performing certain tasks with an eye toward viewing everything as a system.
- Use rapid iterations of prototyping and testing: refine designs quickly and effectively based on actual user feedback.

Similarities and differences between design thinking and human-centered design
First off, design thinking and HCD have some similarities:

- Focus on empathy
- User-centered
- Iterative

Although both design thinking and human-centered design are implemented to develop user-centered products, design thinking is about creating and developing new products quickly, while HCD happens *after* product design and focuses on improving the user experience.

- Design thinking considers people in general, while HCD breaks the audience into segments and focuses on smaller issues.
- Design thinking focuses on creativity, while HCD focuses more on data.
- Design thinking works toward innovation, while HCD prioritizes good usability and user-experience.[4]

---

[4] "Usability vs User Experience: What is The Difference?" (Grigoryan 2024)

Combining design thinking with human-centered design

Design thinking and human-centered design have many similarities as well as differences, yet their similarities make them easy to combine to create even more innovative, user-friendly products. Starting with design thinking, which focuses on people in general and innovation, then following up with HCD, which focuses on audience segments, data, and usability, can help you create products that add value for your audience.

Double-diamond content design process

Drawing on both human-centered design and design thinking, the double diamond came to life. The British Design Council originated the Double-Diamond design process in 2005.[5] The model, shown in Figure 2.3, helps designers and creative people follow a structured approach to problem solving and deliver projects efficiently. Through my experience designing and delivering content, I came up with an approach, based on the double diamond, that has worked for me.

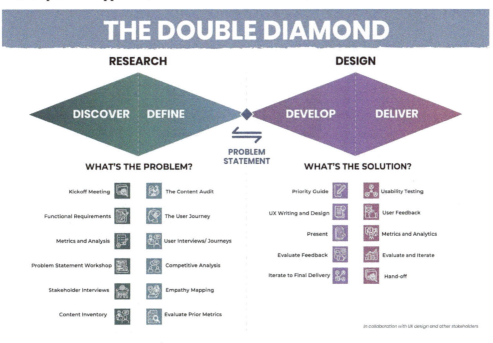

Figure 2.3 – Double Diamond[6]

---

[5] "The Double Diamond" (British Design Council)

[6] Image derived from The Double Diamond, which is © Design Council under the CC by 4.0 License [https://creativecommons.org/licenses/by/4.0/]

The tools, shown beneath each phase in Figure 2.3, are not all necessary or required, but they are tools you can use throughout the process, depending on the project goals, timelines, and budget. These tools can yield invaluable insights into the user, your stakeholders, and the product.

Selecting tools for each project phase

How do you identify which tools will give you the critical information about the user that you're looking for? Many factors go into selecting the right tools. The selection process happens in the project kickoff meeting. You want to choose tools that are:

- High impact
- Most value-add
- Timeline appropriate
- Cost-effective
- Within scope

Discover: When deciding what tools to use or tasks to perform, look at which will have the most impact on naming the problem you're trying to solve. What do you need to know about the user that's going to make the most difference in designing a best-in-class digital product.

The discover phase gives you a handle on the project itself—most importantly the problem statement. Be sure to allow time for this important phase. Once you get a handle on the scope of the work, you can create your schedule.

Define: The define phase requires the most diligent selection process. There are many options for you to choose from to get the most and best information about the user, your business requirements, and your product. Your decision will depend on the time you have, the amount of users available for interviews and other research, and your familiarity with the particular tools. There's no reason to use every tool in the define phase. The tools listed in Figure 2.3 serve as a resource of methodologies available to you for research.

Once you complete the discover and define phases, combine your findings and revise the initial problem statement, so that you're sure you're solving the right problem at the right time.

Design: In the design phase, the tasks you choose will depend on the problem you're trying to solve and all that you've discovered about content in the define phase. Sketching out designs using only content is a fascinating exercise, which you can follow up by bringing those sketches to the

UX designer and working together using content for guidelines. This is where you see, through iteration, the power of content in design.

You also get to work with developers, CMS professionals, and other stakeholders, such as marketing, legal, and UX research, to get final approvals on your designs before you move forward into the deliver phase.

Deliver: The tools in this phase are critical for evaluating the success and efficacy of your product. For example, in one of my jobs as a manager for a high-visibility, national-impact release, we monitored metrics hourly and made updates we agreed would be most useful to the user. Usability testing and user feedback were also critical for the success of this project.

You'll work with developers and CMS developers to understand the feasibility and efficacy of your designs, a conversation you be having along the way.

Of course, all of these decisions are related to time and budget, so bear that in mind when deciding which tools to use.

Double-diamond content-design process

Your content-design process will be unique to your organization and to the way your teams work together. You may have a different processes for different types of projects. Some projects may require a longer and more involved approach, while others are lean and mean for quick-turnaround requests.

Define the process

To define a content-first design process, get feedback from UX and external stakeholders. My approach in the TIAA example I described at the beginning of this chapter was to get feedback from my content-design team, then workshop with UX and other stakeholders to construct the most effective flow based on the double-diamond approach.

Iterate

Once your team has the process in hand, you can iterate with all involved to make sure that the process accurately reflects the needs and experiences of the team.

Educate

Lastly, the team shares the process across the UX organization as a means of both getting feedback and educating the organization about how this resource will be used to guide the project and how you will also use it as a process prototype for future projects.

As I indicated earlier, smaller projects may require much less definition and less discovery up front. Just be sure to document the steps you identify for each project and have your content-design process documentation handy for project kickoffs. You never know when a small project might balloon into a much larger engagement.

While your content-first design process will vary depending on your organization, UX design methodology, and how your teams work together, setting up a document that people have agreed to use as a model for future projects will, over time, help you effectively manage those projects and deliver best-in-class digital experiences. This process will continue to evolve as teams learn more about the benefits of content-first design and open up to getting content involved early.

CHAPTER 3

The discover phase

During the discover phase of the content-first design process, you work to gain insight into the user problem. What issues are users having and how do you want to go about solving them? Here are the components of the discover phase covered in the next few chapters:

- Project kickoff
- Stakeholder interviews
- Problem statement
- Content inventory

While having content lead the process is ideal, it can be a challenge even getting a seat at the project kickoff meeting. So how do you get involved as early as possible?

Here are some of the challenges you may run into:

- Getting a seat at the table early in the process, preferably project kick-off
- Getting stakeholders on board—from designers to business partners
- Getting the time and budget to conduct the work
- Getting stakeholders' to take the time to listen to your ideas

You can try some of these methods to improve your chances:

- Emphasize that this process is a collaboration, not a takeover by content
- Educate about the value content brings to the table when conducting content-first design
- Address your colleagues concerns about implementing this process
- Prove the value of content-first design through measurements, such as testing, analyzing completion rates, and other metrics

I discuss these methods in detail in Chapter 12 and Chapter 13, as well as in this chapter.

The project kickoff

To get started, plan the project kickoff meeting. Many of the content designers I've talked with have had trouble getting the people they work with in UX and other key business and legal stakeholders to let the content team lead the kickoff meeting. Let's assume for our purposes here, that you are able to run the meeting. If not, find ways to bring content design to the table as early in the product design process as possible.

From the very first meeting, a content designer needs to be there and, preferably, to run the meeting. From user research to evaluation and analysis, design, and testing, a content designer who knows how to create user-centered content is critical to the success of a project.

Your presence at the table helps you build relationships with stakeholders and allows you to demonstrate the value of content design in helping achieve success. By the end of the project, the team will realize that bringing content design in early is key to the success of the project.

Challenges to getting everyone on board

Sometimes key stakeholders resist participating in project meetings with content designers because either they don't know the benefits of working with content designers or they don't understand how to engage with a UX team.

If stakeholders don't want to make themselves available, take the time to help them understand the benefits of working together, either through stories of previous project successes, metrics regarding site improvements achieved through a solid content design, or one-on-one meetings where you can talk them through their issues and concerns. Seek to understand your colleagues, their concerns, pressures, and reservations, so that you can address each issue thoughtfully. Careful listening is the key to this and any other conversation in the design process.

Convincing stakeholders of the value of content-first design

Challenges getting your organization on board with content-first design can happen at many levels. If your business partners aren't educated about the role content design plays, you have some work to do in getting them up to speed. They'll want to see proof in metrics that translate to dollars and cents. Another challenge can be getting stakeholders to take time from their schedules—they're less likely to find the time if they don't understand the benefits of content-first design.

You may find that some of your stakeholders—product managers, project managers, business advisors, UX designers, and even UX designers on your team—don't understand the value of content design, and you need to educate them. Many don't understand the difference between marketing copy, which sells, and user content, which helps people interact with the product.

But it's hard to show the value of good content-first design without first doing some content-first design. How do you approach this? One way is to find something small that you can fix and measure to substantiate your efforts with metrics and research data.

For example, at TIAA, customers were abandoning withdrawal transactions early on in the process. We found that if we added an initial screen that let customers know what to expect, how long the process would take, and what they'd need in order to get through it, the drop-off rate went down, calls to the call center went down, and the online completion rate went up.

When you show measurable results, it makes an impact on the people you want to convince of the significant role content designers play.

Getting stakeholders on board

So, given potential resistance, how do you get your stakeholders on board? Let's start with this: who are the stakeholders? They can be some or all of the following:

- The product and business teams working on the metrics and business goals for the project
- The UX team, including content designers, UX designers, the usability team, user research, and the accessibility team
- Legal or compliance representatives
- Marketing
- Anyone else who has a role in the project or will be affected by it

Be sure to identify the stakeholders specific to your project. Make a list of everyone who will have a hand in making your project a success. As you learn more about each stakeholder, this document will evolve and grow and become something you can refer to in future projects.

Remember, the one thing that the whole team has in common—project success. This is the most important thing to keep in mind when interfacing with your stakeholders. You all have the same goal. You may have different ideas of how to achieve that goal and different skills to contribute, but you each want the project to succeed. This gives you some common turf, so that when you're listening to each other and disagreeing and wagging your head, you can reel yourself in and remember, *We all want the same thing.*

This can open you up to listening more to what other team members care about and what knowledge they have. You never know what you're going to learn when you simply *listen.*

How to run a project kickoff meeting

During this meeting, your project team will share a series of documents containing background information, such as existing metrics and user research, business goals, budgets, timelines, top use cases, and a statement of how you will measure success. At this point it's too soon to articulate the problem statement and the highpoints of the double diamond: define, discover, design, and deliver. First, you must gather the information that already exists, understand the history of the functionality that's causing the pain point at hand, meet the team, and begin to build the relationships that will help you succeed.

The project kickoff meeting is also where you can introduce the double-diamond content-first design process. Here's where sending out a meeting agenda can save time and set expectations. This helps everyone understand what they need to know, what they need to bring, and where they can offer feedback. Some will decide to skip the meeting, and you can deal with that when you get meeting decline notices. If the person is critical to the meeting, reach out to them and express why their presence is important.

You're agenda might look something like this:

- Introductions
- Project scope
- Define process items
- Responsibilities
- Meeting schedule and communication touchpoints
- Success metrics or other measurements
- Next steps

Introductions: Here's your chance to build relationships with your team and get to know people you may be working with. Building relationships across the organization is critical to the success of your work and the content design department. Take time to gain an understanding of each person's role in the project, their concerns, and how best to communicate with them.

Project scope: Understanding the boundaries of the project can save time and money. You don't want to do work that hasn't been specified, even if you see work that needs to be done. If you see some out-of-scope work that needs to be done, tell your project lead what you've found, let them know how the work can enhance the project, and work with them to convince stakeholders to add it to the project scope. Additions to project scope are about more than spending extra time; they are also about spending extra money, so be sure to respect your project scope.

Work process: Here's where you can share the work you've done to create the content-first design process and get buy-in, if you don't already have it. If you've used the content-first process with these stakeholders before, this step is easy. It becomes about how you'll work with team members and any further expectations or questions about the process, such as how to deal with changes in scope or deadlines. If you don't have buy-in, yet, you have some work to do. See Chapter 11 for ideas on how to get buy-in.

Responsibilities: Outlining who is responsible for what can save a lot of time and misunderstandings, as well as hold team members accountable for what's on their plate. Since you've developed your work process, you can easily show your responsibilities. Be sure to capture everyone else's role and responsibilities in the project notes, too. This will be important once you're at work so that deadlines aren't missed and misunderstandings don't occur. You can review progress on goals at team stand up meetings.

Meeting schedule and communication touchpoints. The cadence of meetings will set you up for success. If you can establish what's expected at each meeting, even better. For instance, is the meeting an internal review with the design team or a stakeholder review? What work should be ready to show at each meeting. Meetings are an important touchpoint, as are regular communications among team members. Defining these elements will help you prepare and use meeting time effectively and efficiently.

Success metrics or other measurements. Everyone on the team needs to know what the success of the project will look like. Are the flows measured with metrics such as how many people complete a purchase on an e-commerce site or how long people stay on a page? What improvements are you looking for? Understanding these and other questions will help you gain insight into the problem you identify in your problem statement workshop and steer you toward success.

Next steps. While you've established a few next steps by setting up a communication plan for the team and a meeting cadence, what other next steps have you agreed to take? Capture these in a document and email it to the team with all of your notes from the kickoff meeting.

TIDAL BANK

Case Study: Project kickoff

The design team at Tidal Bank reviews the double diamond content first design approach to UX design. This framework helps the team understand how to approach the design problem by thinking through the Design, Discover, Develop, and Delivery phases.

The team reviews the diamond together and thinks through all of the possible tasks and exercises they would potentially do during each phase. They will use the double diamond during the project kickoff to help them select the activities they will complete during the project.

The team will also regularly refer to the double diamond throughout the design process to make sure they are still on track and doing tasks that make sense within the double-diamond framework.

To start the project, the team sets up a project kickoff meeting. Anne works with Roberta, the UX design lead, and John, the product manager, to host the kickoff with the full team. The kickoff meeting is key to a successful project, as it creates alignment among members of the project team and helps define what tasks will be carried out in order to create the best digital product.

This group includes the UX design team, the product and business teams, legal and compliance representatives, marketing team members, and anyone else who may have or need insight into the project. An agenda is sent out ahead of time so that expectations of the meeting are clear.

The first order of business is to discuss the current issues with the Tidal Bank website. They learn from the analytics team that the time users spend on the website isn't translating into new accounts. They tend to drop off in the process at key points that need to be addressed.

These include:

- During the account exploration process
- When deciding which type of account they want to open
- When it's time to enter bank information and fund the account
- Before the final **Open account** button

The product team notes that there are a lot of places where users bounce from the page. The completion rates are very low across the board for a variety of accounts. The product team also notes that there is a relatively high volume of calls coming in from users who are confused by the website, as well as long lines at bank branches where users are coming in to complete simple tasks that could be accomplished online.

The product team would like to see a clear improvement in conversion rates for opening accounts, a decrease in call volumes for simple tasks, and fewer in-person branch visits for tasks that can be completed on the website. They want the design team to review the website and come up with new designs to help address these issues.

Now that there's a better sense of the problems the team is tasked to solve, the team takes the time to discuss how they might approach them. Individuals start throwing out ideas for steps that they think the group should take, and Anne documents them on a white board.

Some of the ideas the group explores include: content inventory, user research, competitive analysis, and priority guides.[1]

Priority guides provide an outline of information on a web page, sorted by relevance, without layout specifications. "Design Process In The Responsive Age" (Clemens 2012)

A content inventory will be useful because it will take the content designer through the entire website experience and information architecture. The group will also leverage the inventory when it comes time for the content audit.

The product team has emphasized the need for actual feedback from users to accompany the metrics they already have from the website. The team suggests user interviews during the define phase.

The team needs to explore why these areas pose pain points, so they plan to conduct the following user research:

- Empathy mapping
- User interviews
- Testing prototypes

The team also decides they want to better understand what their competitors are doing and how their competitors are solving some of these same questions. So they decide to do a competitive analysis to see what is working and what opportunities they might have to create a better product than their competition.

After discussing a variety of possible activities they could use to create a better user experience, the team settles on a dozen they feel are the most important.

Based on time, budget, and people resources, the team decides to take on the following activities from the double diamond.

Discover

- Problem statement workshop
- Content inventory

Define

- Content audit
- Competitive analysis
- User research & empathy mapping

Design

- Priority guide
- Prototype starting with content
- Stakeholder presentation
- Testing with real users
- Iterate on designs

Develop

- Hand off to development
- Iterate to improve the product

Why choose each of these discover-phase tasks?

Problem statement workshop: The first step in any project is understanding the problem you're trying to solve. The team wants to make sure that they solve the right problem for the user.

Content inventory: The inventory immerses the team in the current information architecture and user experience. It also sets them up with a worksheet for the content audit.

Content audit: The audit takes the team through the user experience again, forcing them to consider and rate everything on the page according to heuristics the team has developed.

Competitive analysis: The team wants to understand how Tidal Bank's competitors work to get customers to open accounts, so they plan to evaluate and compare their competitors' sites to identify strengths and weaknesses.

User research and empathy mapping: User research is essential because it is impossible to build an experience that meets user needs without hearing directly from users. The team decides to interview users and conduct an empathy mapping exercise with users to better understand their needs and pain points.

Priority guides: The team decides to build priority guides before they start the design phase to make sure that they organize information in clear and useful ways. This will ensure that the designs they create have all the information they've identified that users want.

Prototype: A prototype will give the team something tangible to respond to with constructive feedback. The team might go from low-fidelity to high-fidelity prototypes as they hone in on a solution.

Present to stakeholders: To get consensus on content and potential designs, the team will present their findings and early sketches to all of the Tidal Bank stakeholders for feedback. By walking stakeholders through their designs and design decisions they can create consensus.

Test designs with users: Once the team has designed prototypes that stakeholders are on board with, they want to test with users to make sure the designs actually solve the problems they hope to solve. The feedback they get from users at this stage is essential to ensuring a successful design.

Iterate on designs: Once the team has feedback from stakeholders and users, they will make updates and changes to their design to better reflect that feedback. No one gets it exactly right the first time, so iteration is an important part of the process. They will also submit designs to compliance or legal to ensure that they keep everything in line with any legal requirements.

Hand off to development: Once the designs are complete, the team will hand them off to the development team to build. The design team will stay in close contact with the development team during the process to make sure that the designs are built correctly and to ensure quality.

Iterate to improve the product: Finally, once the designs are live in production and available for users to interact with, the team will regularly review metrics and user feedback to see if the changes they've made work as expected and if there are additional ways to improve the product. The double-diamond design process never truly ends. User pain points will crop up in the testing, and you may go back to tools from the discover, define, design, and delivery phases.

CHAPTER 4

The problem statement

The first thing you need when entering a project is an initial problem statement—what are you trying to solve for the user and the business? Are there any assumptions or hypotheses that can inform your research? A problem statement should include the problem to be solved, who the users are that will be affected, and what the impact is on the organization.

Oftentimes, content will receive a problem statement with a solution attached or, even worse, no problem statement, only a solution. For example, the product designer comes to the meeting and says, "We need to do a, b, and c based on our research metrics and business needs." This may be great at first glance, but it may fall apart under closer consideration. Metrics show the *what* but not the *how* or *why*. Business requirements are for the business, not for the user.

Content-first exploration provides a deeper level of information and understanding about pain points, which allows you to offer a solution that really works for the people using the product. Without a solid content point-of-view based on user interviews, empathy maps, priority maps, and other content-based research designed to understand the user in a deep way, a design will not succeed, no matter what backup information is presented. A solution that doesn't follow from a good problem statement will likely not succeed.

When stakeholders provide backup information, you can tell them that you've conducted extensive user interviews. Let them know that you can go a step deeper and build a problem statement based on content-first design and using some or all of the techniques you can show them in the double-diamond diagram.

Be sure when you start to write the problem statement that you introduce a content-first point-of-view. Note that it will evolve as you continue your research. This could require a fair amount of work, depending on your team and the project, but the outcome will make this initial step worth the trouble.

> **An example problem statement**
>
> Users trying to transfer money from an IRA account to an annuity are dropping off the digital experience. We have metrics on where they're dropping off. This is a problem for the company because the call center is inundated with customer questions, costing the company millions of dollars a year, and we don't want people to move their money to another financial services firm. Users are frustrated because they want a simple solution through the web or app to move their money without having to talk to a representative or fumble around with an experience that doesn't work for them.

With a problem statement firmly articulated, you can work to understand users on a deeper level through a define phase. The more you can get into the user's head and their emotional experience, the more you will be able to meet them where they are and meet their needs with a seamless digital experience. Chapter 5 and Chapter 8 delve into the details of how to conduct this research.

Problem statement workshop

The purpose of a problem statement workshop is to collaborate with your team and stakeholders to define the core problem you're trying to solve, why it matters to users, and what success will look like. It's an opportunity to begin a collaborative effort with the team and build solid relationships, as well as gather existing metrics, constraints, user research, and other information the team has to offer. Anything that helps you know your user better will be relevant to this workshop.

Conducting the workshop takes some planning and will put your facilitation skills to good use. You'll need to keep people engaged and productive. Here are some ideas to get you started.

Invite the right participants

The participants should include everyone on the project: UX, content, product, and other business stakeholders who have the right experience and knowledge to craft an effective problem statement. It's important to send out an agenda that outlines the purpose and the flow of the workshop, as well as the proposed outcomes. You may even want to set up a pre-meeting with those hard-to-wrangle people who are often too busy. Convince them of the critical nature of the workshop and review the agenda with them to get them onboard.

Establish rules of engagement

Be sure to create a set of ground rules that will keep the workshop collaborative and on track. People need to listen closely without interrupting, ask important questions, and keep the conversation focused on the task at hand. To get started, it may help to have an icebreaker. Perhaps go around the room, or virtual room, and have each person introduce themselves, their role, and one fact about themselves that nobody knows. That's just one idea. Be creative and get the group into a team spirit.

Provide a workshop structure

A solid structure for the workshop will keep people engaged and focused. You may consider something like this: the problem, the user, the impact on the organization, and the impact on users. This provides great fodder for group discussion where everyone can share ideas and any evidence they have to challenge assumptions. Capture these ideas on sticky notes or with a tool such as miro.[1]

TIDAL BANK

Case Study: Problem statement workshop

The Tidal Bank team needs to create, as a team, the problem statement for Tidal Bank's onboarding project. Anne, the content designer, decides to hold a problem-statement workshop for the onboarding project.

She invites:

- Herself, the content designer, workshop facilitator
- Roberta, UX design lead
- Amy, UX designer
- Santos, UX designer

---

[1] https://miro.com

- Millie, UX researcher
- Vamshi, product manager
- Kerry, sr. product manager
- Felicity, business partner
- Arez, marketing assistant

Anne, sends all of the invitees a clear agenda ahead of time, outlining the plan for the two-hour workshop.

Agenda:

☐ Introduction

☐ Establish rules of engagement

☐ Icebreaker

☐ Small group project

☐ Large group readout

☐ Large group brainstorming

☐ Refine problem statement

☐ Wrap up

Anne starts the meeting with a short introduction of the reason for the workshop and a walkthrough of the agenda. She also outlines general rules for how the group should collaborate together.

Rules of engagement:

- Send an agenda with a list of meeting goals and attendees.
- Let people know that kind of feedback the team is looking for so no one is distracted by other things they might see.
- Listen until people are done speaking, then repeat what you heard to make sure everyone was heard correctly.
- Work to create consensus in the room, while working for what we know is best for the customer.

- Identify next steps for the team to take before the meeting ends.

- Send follow-up notes with next steps, next meeting time, and who will attend.

Anne splits the group up into groups of three for the first exercise. She likes to start with smaller group exercises first, because some people feel less intimidated and find it easier to share in smaller groups.

Then Anne explains the first exercise. She asks each small group to work together to think about what a typical user of the Tidal Bank might look like and what their motivations for using Tidal Bank's website might be. She sets a timer and lets the groups work among themselves for at least 15 minutes.

Once the timer goes off, Anne brings the groups back together. She asks each group to share what they came up with during the small group exercise.

One group talks about a user who is a young adult who has never had a savings account before. The user they imagine is someone who is generally digitally savvy and would prefer to complete tasks online rather than in person. However, this user does not know a lot about financial products and isn't sure what savings account would be right for them.

The small group emphasizes that this user would be quick to use the website if it is easy for them to complete their tasks, but they would also want a lot of explanatory information to ensure they make the right choices for their situation.

After each group shares, Anne leads the group through another exercise where they put sticky notes on the white board in two sections. One section is for characteristics of possible Tidal Bank users and the other is for user needs.

The group spends another 10 minutes on this exercise. Once everyone has put sticky notes on the board. Anne leads the group through a discussion of what's on the board. She starts grouping items that are similar and notes any crossover.

From this exercise, Anne starts to see certain patterns and suggests that their problem statement focus on providing information about financial products and making the sign-up process as simple as possible.

The group discusses these ideas and works through a draft problem statement:

As a new Tidal Bank customer, I want to understand what financial products are available to me so I can sign up for the product that meets my financial needs.

After they create the draft problem statement, Anne takes it to a larger group of stakeholders to gather feedback and make sure it resonates with the team. Anne makes small changes to the problem statement based on the feedback and then shares out the new problem statement with the large group, emphasizing that this problem statement will be the basis of the project moving forward:

As a Tidal Bank customer, I want to understand what financial products are available so I can sign up for the product that meets my needs.

Identify assumptions

The problem statement process includes identifying assumptions from the perspectives of both the business and users. Here are some examples:

Assumptions from business:

- I believe my customers need to…
- These needs can be solved with…
- My initial customers are (or will be)…
- The #1 value a customer wants to get out of this project is…

Assumptions about the user:

- Who is the user and how does the deliverable from this project fit into their daily life?
- What problem does the project solve?
- What features are most important?

Brainstorm to identify other assumptions you may have from business and user perspectives. This will help you understand where you're coming from and identify any pain points in the team's collective thinking.

Plan project tasks

What strategies in the define and discover phases will be most effective, have the highest impact, and add the most value? Establish a set of guiding principles to measure this by, such as timeline, budget, and dependencies. Once the team decides which tasks to perform at each stage, sketch out a process that includes a timeline, deliverables, and all stakeholders' roles and responsibilities.

As a content designer, I do a lot of research to gain a strong understanding of whatever product I'm working on. There's nothing worse than coming into a meeting with a designer and not having a functional knowledge of the product. When content designers have a strong understanding, it empowers us to go beyond copy. It allows us to have a foundational point of view on information hierarchy, voice, tone, logic flow of the experience, labels, and gaps in research that would create a better experience.

—Kelcea Barnes, Freelance Content Designer, former Senior Content Designer Upwork, 2022 Meta Design Fellow

TIDAL BANK

Case Study: Create a project plan

Now that the team has a clear idea of the steps they plan to take, they need to figure out a feasible timeline for completing the work and to assign tasks to the team members.

Felicity has timelines from the company leadership about when they want to launch the new pages. Roberta has a sense of how long design tasks needed for the work will probably take.

Felicity and Roberta talk through a rough draft of the plan and outline potential due dates for tasks. For example, Felicity expects gathering the full business requirements for the work to take about two weeks to complete and points out that this needs to happen before design work can commence.

Roberta expects the early discovery and research tasks, including collecting metrics, conducting a content inventory and audit, comparing competitor designs, and exploring user research will take at least six weeks to complete.

Using these basic benchmarks for tasks, the two build a rough outline of the project plan (see Table 4.1). Then they bring the plan to the larger team during the project kickoff to get feedback about how long the team expects tasks to take and if there are tasks that need to be added or removed from the plan.

The team also spends time assigning tasks from the plan to various members of the team so that everyone knows what they'll be working on and when. Once everyone is aligned on the approach the plan is ready for execution.

Table 4.1 – Tidal Bank project plan

Task or Tool	Description or Deliverable	Lead and participants	Due Date
Problem statement workshop	Project problem statement	Anne	8/15
Gather business requirements	Discussion with business partners and outlined in project document	Vamshi, Felicity	9/1
Collect existing research, metrics, and other data	All documentation of metrics and user research gathered in one place for analysis	Roberta, Amy	9/15
Stakeholder interviews	Document with transcribed, edited interviews	Millie, Anne	9/30
Content inventory	Inventory spreadsheet	Anne, Santos	9/30
Content audit	Audit spreadsheet	Anne, Santos	10/15
Competitive analysis	Competitive analysis report	Millie, Anne, Amy	10/30
User research	Research analysis report	Millie, Anne	10/30
UX user journey flows	Journey map	Millie, Anne, Santos	11/7
Priority guide	Content priority outlines	Anne	11/21
Prototype	Figma design file	Roberta, Amy, Santos	12/1
Present to stakeholders	Design presentation	Anne, Roberta, Vamshi	12/5
Iterate on feedback	Figma design files	Santos, Amy, Anne	12/10
Work on final designs and content	Figma design files with final content	Santos, Amy, Anne	12/31

CHAPTER 5

Interviewing stakeholders

Once you have an initial problem statement, you can go a level deeper by talking with stakeholders. Interview stakeholders, or anyone who has an interest in the project, in order to build a solid relationship with them and gather any insights they may have into the project. Do these interviews early in the discover phase to welcome your colleagues on board and to start gathering information they may have to help you and the team deliver a successful outcome.

Get to know anyone involved in the success of your project. These may be other designers, content specialists, accessibility experts, product managers, or business owners. Whoever you interview, keep in mind that the interview is a reflection on you and the project team, and if done well, a good interview can help engage stakeholders, help you understand the broader project beyond content needs, and let stakeholders learn more about your capabilities.

The overall goal of the stakeholder interviews is to improve the design process, team communication, and team engagement. Rather than having an agenda of how you'd like the project to go, be curious, ask questions, and seek consensus.

Get to the heart of the matter

To dig deeper into the conversation with your stakeholder, try asking some of these questions, provided by Nielsen Norman Group.[1]

Tell me more about that?
Can you expand on that?
Can you give me an example?
Can you tell me about the last time you did…?

---

[1] https://www.nngroup.com

To further explore your stakeholders actions and reactions, try these questions, also from Nielsen Norman Group.

How do you feel about that?
Tell me why you felt that way.
Tell me why you did that.
Why is that important to you?
Why does that stand out in your memory?

Benefits of the stakeholder interviews

By getting to know individual stakeholders, their concerns, knowledge, and desired outcomes, you can gain insight into how best to communicate with and work with them, and you can discover the history, context, and knowledge that they bring to the project. This can help you dig up previous solutions that didn't succeed and ask *why*? You can also find other people in the mix who have issues about the work—people that you should know about if you're to navigate the landscape adeptly.

During your stakeholder interviews, make sure you listen and hear what they say. Listen for their vision for the project and work together to align on that vision. Learn their concerns, fears, and worries. What are their deadlines, their boss's demands, and what do they need from you in order to deliver success? What do you need from them?

When people feel heard, they're more likely to listen to you and consider your recommendations and requests. This is where you begin to create alignment on the project vision, anticipate obstacles, and create steps and a cadence of communication to help you achieve success—together.

In addition to helping you build solid relationships, interviews also help you gather:

- Business goals and measurements for success
- Business, user, and technical requirements
- Known constraints
- Existing research and metrics
- Any data related to past project outcomes

Be prepared

When it comes down to the actual interview, compose your questions ahead of time. You may not get to all of them or find all of them relevant once you're in the conversation, but they serve as a roadmap. For instance, if you dig deeper into your colleague's comments, you may take a side road in the conversation, but you'll always have your roadmap to get you back on track.

Let the conversation be fun. Build a rapport that will serve you throughout the project and make the relationship an enjoyable one to the best of your ability.

If you're in a jam, or you can't nail down a time with a stakeholder, you can offer an email interview. However, if you do this, get them to agree to multiple rounds so you can dig into areas of interest. They'll be less likely to answer a lot of questions in an email interview, so narrow it down to a few key questions.

Once you've drawn as much information as you can from your interviews, understanding the insights, defining the outcomes, and communicating that information and your insights to the team becomes the most important part of your interview process.

Stand back and look at the interviews as a whole. Do themes arise that you can explore by reading through the interviews again? Are there common concerns or an emphasis on potential obstacles? Do they agree on how to measure success, deadlines, and team politics? Get answers to any outstanding questions, then outline your findings, and distribute to the team as appropriate via email, a presentation, or a meeting.

CHAPTER 6

Taking inventory

A *content inventory* is "a list of every piece of digital content you currently have, captured at either the page or asset level. It includes specific characteristics about each piece,"[1] while a *content audit* is an evaluation of existing website content contained in the content inventory.

The inventory can include page URL, page title, meta data, pages hits, and additional information that can provide more insight into user engagement. A content inventory is an excellent starting point for your audit as it provides a spreadsheet with every piece of content you want to evaluate. Content inventories can be conducted using a wide range of inventory tools. many of which are available for free. For this book, I use Screaming Frog's SEO Spider tool.[2]

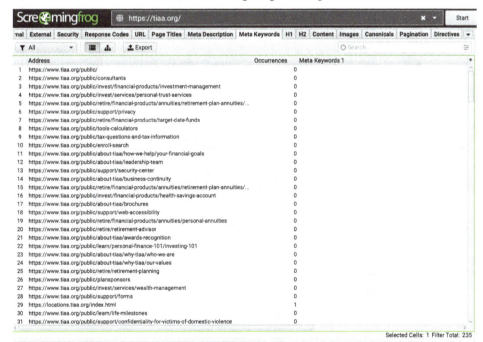

Figure 6.1 – Output from the Screaming Frog SEO Spider tool

---

[1] Nielsen Norman Group

[2] https://www.screamingfrog.co.uk/

When you look at the visual from Screaming Frog (Figure 6.1), you see the list of page URLs and a series of tabs across the top of the sheet that let you explore various attributes of that page, such as headings (H1 and H2) and meta keywords.

A content inventory familiarizes you with how your website is organized, how much content you have, and where content isn't working, which is evident by bounce rates and the amount of time spent on a page. The content audit, which I discuss in Chapter 7, analyses this data further.

For me, content projects are a process of asking and answering questions. No matter how large or small the project, the starting point is knowing what you have to work with—what do I have and how far is that from the target state I'm trying to achieve? An inventory is a foundational step in being able to scope and plan your project and it provides early indicators of patterns that you'll want to look for in an audit.

—Paula Ladenburg Land, Principal Consultant, Enterprise Knowledge, LLC

Why conduct a content inventory?

A content inventory gathers relevant information from a website for your analysis and catalogs the content to help you understand attributes such as where content lives, how old it is, how much time users spend on each page, and how often users immediately leave a page.

You need to know what you have before you redesign a website or work on part of it. The content inventory will help you deepen your understanding of your content and how it all works together. It will help you:

- More accurately scope and budget the project
- Learn the structure of your website
- Get a handle on what content you have and where it lives
- Identify what content you need to keep and what you can get rid of.
- Track progress during development by providing a starting reference
- Gather important information about each page on your website

Research methods or activities are simply ways to answer questions. They are tools in a toolkit. There's no best tool, only the right tool for the job given your needs and constraints. Some tools are simpler to use and some require more expertise.

The decisions you need to make and the gaps in your knowledge determine what type of data you need, and how much and how rigorous. Are you interested in descriptions or measurements? Do you need both to understand something about human behavior and understand the frequency of that behavior to a degree of statistical confidence? That will tell you whether you need quantitative, qualitative, or mixed methods.

The fundamental tool is the question itself. If you don't have a good research question (not interview question), nothing else matters. If you have a good one, everything else is a lot more likely to fall into place, including the method, analysis, and insights.

Once you're clear on your question, then it's a matter of time, budget, and expertise. If you need to make a decision tomorrow, the best thing you can do is spend today reading the internet. There is a lot out there—not just existing research findings, but also reporting and expert opinion, and the tried and true "see what people are doing".

—Erika Hall, Author of *Just Enough Research* (Hall 2024)

Creating the inventory

You can either create your inventory by hand, if there is time and the scope isn't huge, or you can use a crawling tool, such as Screaming Frog, or an export from your CMS.

Manually creating the inventory

By manually going through and capturing attributes of each page, you get the opportunity to dive into your content across the site or project area. This can be a time-consuming, arduous task, or a mind-expanding ride on the magic bus. You never know what you will find, so buckle up.

To create the inventory manually, create a spreadsheet with the fields you want to include. Here are a few suggestions:

- URL
- Page title
- Meta description
- Meta keywords
- Headings (H1, H2, etc.)
- Page views
- Bounce rate
- Time on page
- Other available analytics

These are only suggestions for a basic inventory; depending on your goals, these fields will change. For example, if you're doing an inventory to audit the page statistics of a website or part of a website, you'll want more page analytics (page views, bounce rate, etc.). If you're looking at search-engine optimization (SEO), then you'll want to dig in more to the metadata so you can reconcile your findings with what's on each page.

Creating an inventory with a crawling tool

For the purposes of this book, I use the Screaming Frog SEO Spider as the crawling tool. There is a free version online with a video demonstration on how to create your content inventory. The free version (screamingfrog.co.uk/seo-spider) is limited in how many URLs you can crawl, but it is useful for evaluation.

Content inventories typically precede content audits, because the inventory provides a template for your content analysis. Once you have a spreadsheet with all the content listed you can add columns for the various attributes you want to analyze, such as findability, readability, and usability, as well as compliance with brand and style guidelines.

TIDAL BANK

Case Study: Content inventory

Anne and Santos start working on a content inventory by reviewing the current pages on the Tidal Bank website and making a spreadsheet to collect their data. They use the Screaming Frog SEO Spider and Google Analytics to get additional data about visits to the site, including which pages users visit most often and stay on the longest.

The spreadsheet documents all of the information they plan to gather, including:

- Page URL
- Page title
- Page description
- The first Heading 1 on the page
- The first Heading 2 on the page
- The number of views the page has had in the last 7 days
- The bounce rate (the percentage of visitors who immediately leave the page)

For example, through the Tidal Bank Google Analytics account they see that the Accounts page has a bounce rate of 53.30% while also being one of the most viewed pages on the site (line one in Table 6.1). This implies that although people are regularly navigating to the Accounts page, they don't seem to be finding the information they are looking for.

Table 6.1 – Tidal Bank content inventory

URL	Page Title	Page Description	H1	H2	7-Day Views	Bounce Rate
/accounts	Accounts	Tidal Bank offers convenient account services to make banking and life easier. Learn more.	Check-ing Accounts		15,257	53.30%
/checking	Checking	Tidal Bank offers our members free checking with rewards. It's easy to switch and start saving money today.	Check-ing	Mobile banking	7,634	33.40%
/saving	Saving	Tidal Bank offers a wide variety of products to help your business save more money. Open an account today!	Savings	Holiday club	7,469	36.40%
/certificate-of-deposit	Certificate of Deposit	<none>	Certific-ate of Deposit (CD)	What are CDs?	5,251	14.30%
/money-market	Money Market	<none>	Money Market	What account is right for you?	5,415	42.90%
/what-are-cds	What are Certificates of Deposit	Tidal Bank offers a wide variety of products to help your business save more money. Open an account today!	What are Certific-ates of Deposit?	What account is right for you?	746	89.30%

URL	Page Title	Page Description	H1	H2	7-Day Views	Bounce Rate
/CD-FAQs	CD FAQs	Learn how much you can save when you use Tidal Bank. Explore our competitive mortgage rates, loan rates, savings rates, and certificates and IRA rates.	CD FAQs	What is a certificate of deposit	823	68.20%
/What-account-is-right	Accounts	Learn more about how you can better manage your finances and prepare for the future.	What account is right for you?	Open a checking account if...	1,983	20.70%
/contact-us	Contact Us	Learn more about how much you can save with Tidal Bank! Find your local branch location and hours, and more.	Contact us	We'd love to hear from you!	11,378	25.40%
/rates	Rates	*Same text as /CD-FAQs*	Check out all of our rates	Account	8,932	48.20%

Table 6.1 shows the results for a sampling of the top-level pages in the inventory created by Anne and Santos.

This content inventory helps Anne and Santos understand some basic statistics about the Tidal bank site. They can see which pages users spend the most time on and which pages users bounce off of quickly. They can also see which pages users find through searches and which pages might not be getting seen by users.

They also notice that users often land on the Certificates of Deposit page via search. This page has a large amount of traffic for the site and a very low bounce rate. However, two of the pages linked to from the Certificates of Deposit page—

the What is a CD? and CD FAQs pages—have much lower traffic and high bounce rates. They make a note of this and plan to follow up on what's working and not working on these pages during their content audit.

By reviewing the information in the inventory, Santos and Anne begin to see an outline of what's working and what's not for the Tidal Bank website. They will use this information to to focus their efforts during the content audit.

CHAPTER 7

The define phase

During the define phase of the content-first design process, you work to gain insight into the user problem. Who are your users, what issues are they having, and how do you want to go about solving those issues? What are you trying to find out and do you have a specific research question? Is there a particular pain point you want to learn more about?

Here are the components of the define phase, which I cover in the next few chapters:

- Content audit
- User journeys
- User interviews
- Competitive analysis
- Empathy mapping

What is a content audit?

A content audit looks at your existing content and assesses its strengths and weaknesses. It is a *qualitative* evaluation that analyzes content against criteria such as your brand and style guidelines, readability, usability, findability, key performance indicators (KPI's), business goals, and other metrics you may want to consider, depending on your project and what you want to learn from the audit.

The starting point for a successful audit is a thorough content inventory, which I covered in Chapter 6. The content inventory provides a template for the audit. By adding columns with content attributes you'd like to analyze, you'll have a complete sheet to work from.

Additionally, the audit will help you scope the project. You might start with the idea that you need to do content changes on just ten pages, but as you go through the audit, you find that there is additional work needed to create a more meaningful customer experience. Or perhaps you decide to stick with the initial ten pages, due to cost, timing, staffing, or other necessities, and come back to the additional work in a later iteration. Using the content audit to help scope your project can prevent you from missing opportunities or overspending.

With your inventory set up as a template for the audit, you're ready to dig into a deeper analysis of what you have on your website. Some inventories and audits will be of a whole website, and some will be smaller, touching only part of a website.

On a small project, say a project that touches only a couple of pages, you may want to inventory and audit the pages preceding and following your project. This will help you understand where the user is coming from and where they are going. You need to know what information the user already has so you can evaluate whether the pages you are auditing fit smoothly in the progressive disclosure of information. You need to evaluate the preceding and following pages to ensure consistency in qualities such as readability and brand expression.

So why do a content audit?

You do a content audit to improve the user experience and help meet business goals. The first thing the content audit allows you to do is to take stock of your current content and assess how it holds up against business goals, brand and style guidelines, usability standards, and so on. You can also identify gaps where content is missing or not making sense.

You can conduct content audits at different points in your project—whether it's a website redesign, a refresh, or a new feature being adding—to assess progress.

Taking the time to do a thorough audit can help you regain control of content that may have become, over the years, a sprawling mess. Even if it isn't a mess, you need to understand what you've got and how it affects both the users and the business.

A content audit will also help you if you take over another content designer's role and need to get an understanding of the site and how it measures up. An audit will also help you understand where you can relieve pain points and make an impact for the business and users. Even small wins gained through an audit can get you a long way in customer experience and in your career.

With the audit in hand, supported by data, you can help stakeholders make informed decisions rather than have them rely on their own opinions. Opinions are the devil's playground in UX. Everybody's got one, and no single one is right. Finding the intersection between business, content, and user needs is a science, and like any other science, it relies on facts, data, and information gathered in controlled environments, such as a usability lab.

The audit comes way before usability studies, but it is a kind of usability study. You can assess attributes such as readability, findability, usability, and so on, which can give you ideas about how to conduct usability studies—what to look for and what to not worry about. The data in the audit will support your website projects over time and offer a ready document to support important decisions related to features and content.

When I worked at TIAA as a lead content designer, we were busy doing a total re-brand of the website and addressing transactions with high drop-off rates. The first part of the job was addressing those transactions, which were a failure for the user, and, as a result, a failure for the company, costing million of dollars a year in customer support calls. My approach was to conduct a content audit analyzing the bounce rates and locating where users dropped; I then used that data to figure out how to remedy those pain points.

What type of audit should you do?

There are so many types of content audits, including auditing for brand and messaging, accessibility, legal issues, business strategy, and more. Here, I will walk you through the essentials of a qualitative audit with a best-practices assessment. We will look at content from the user's point of view and measure it against content-design best practices, company brand and style standards, and what we know about user needs, thoughts, and feelings. Other factors you can analyze include: is the content useful, usable, enjoyable, persuasive, findable, readable, and searchable?

While there are many types of content audits to consider, for the purposes of this book, I will focus on the qualitative audit. There are three different methods for doing a qualitative content audit. Which one you choose will depend on your needs.

Focused audit: A focused audit hones in on a select area of the website. For instance, if you're working on a page that relies on information from another page, you should include both pages, as well as the pages before and after. It's important to understand the user journey. The pages around the one you're looking at may have important information about the user journey, such as what the user needs to know before reaching the page you are auditing. You may also find inconsistencies across the pages and have to decide how to restore consistency.

Complete audit: A complete audit looks at an entire website and analyzes each page. This can be incredibly informative. You can find content that is repetitive, inaccurate, irrelevant, and so on. Getting the big picture view of your site is a great opportunity to educate you and your team about where your content is working and where it's not.

Rolling audit: A rolling audit keeps you auditing over time. You audit sections of your site following a timeline based on your business goals and content concerns. Perhaps a particular page or user flow is not performing well. Then you can audit those pages and the pages before and after them in the user flow to find where the content may be falling short. Audits take time, so a rolling audit can save time and resources by honing in on where the analysis is most needed. This can give you enough information to make improvements without the expense of a complete audit.

Work with stakeholders before the audit

Your stakeholders may be able to give you a variety of resources. For example, product owners and product managers may have metrics that you should track in the audit. Managers and other stakeholders can let you know about their pain points in the project and the pages you're auditing and their goals for the project. Stakeholders may have developed user personas and user stories that can help guide the audit.

When you talk with stakeholders who are outside your discipline, it's important to understand their concerns and needs as well as what they bring to the table. When you understand the person you're communicating with, you can collaborate better and achieve a better outcome.

User personas

User personas are characters designed to represent potential users of a website or app. They're composites of users based on demographics such as income, education, life stage, and career, as well as metrics from actual users on your website. They put into context the range of goals and pain points for a broad audience of users.

Personas can bridge the gap between the organization and its users by offering a window into actual behavior and end goals. Since there are many types of users interacting with your site or app, creating personas can help you sketch out the scope of these people, then design for particular user stories and potential pain points.

User stories

A user story describes a feature that will be added or removed from the product to create a better experience for the user. User stories are written from the user's point of view, and we as content practitioners use them to help guide our work.

For example, a simple story may start like this:

"As a *[user]*, I want *[goal or action]* so that *[outcome or reason]*."

Once you have as much information as you can gather from stakeholders, work together to decide whether or not to audit the whole site or only the parts of it that are causing drop-off or a poor user experience.

As you work with stakeholders to decide what content to audit, guide the conversation toward understanding the particular goals of the audit. What do you collectively want to learn by undertaking this work? Once you identify that, you're ready to go.

Define the attributes to analyze

Your attributes will depend on what you want to learn from your audit. For the purposes of this book, we'll be looking at a qualitative audit of how the content on a website is working. Here are some suggested attributes to consider:

- Accessible
- Inclusive
- Usable
- Readable
- Findable
- Relevant
- Accurate
- Trustworthy
- Timely

- Clear
- Necessary
- Appropriate tone
- On brand
- Business value

In your inventory spreadsheet, create a column for each attribute. You will analyze each piece of content through the lens of those attributes. For example, when you look at a button that will perform an action, you will evaluate whether it is usable, trustworthy, accessible, etc. Depending on what you want to learn, you may identify additional attributes that can help you identify the cause of pain points that keep users from carrying out tasks.

Establish a rating scale

A typical rating scale ranges from 1–5 with 5 being excellent and 1 being awful. Comparing ratings helps you determine whether to keep or delete a piece of content. Creating and then reviewing the ratings will also help you gain an in-depth understanding of the site along with its shortcomings and strengths.

Get to work

You may need to build a team for larger projects, whereas for smaller projects, one person may be able to get the job done. One option is to pair up with someone who knows the subject matter well so that you can analyze the accuracy of the content together.

As you consider each piece of content and the overall content design, think about how each piece fits into the user flow and rate the content based on the attributes and scale you've established. Delegate sections that are out of your area of expertise or to balance your workload. And take your time! This is not a sprint; it's a marathon, and every step counts if you want to understand how the content on the site is working.

Add a column for notes. Here's where you can write down any improvements that come to mind, thoughts about resources to fix a problem, or changes to the UI/UX that could improve business outcomes or the user experience.

User interface (UI)

The UI, or user interface, consists of the visual elements on the page, some of which will take an action, such as a button that says **Cancel** or **Exit**. The UI is the structure through which the content talks to the user.

Analyze the results

With your audit in hand, the next step is to take a good, hard look at what you've discovered. The audit will reveal what the site does right and where improvements, or even a complete change in direction, will make a difference. Do you notice patterns, for example, clusters of similar numbers? Analyze both horizontally and vertically. For example, under usability are there scores that are similar, and what do those scores tell you about the usability of your site? If a problem occurs in one area, you can plan for a small project to make improvements, but if it is pervasive across the site, you've got a bigger task on your hands.

As you go through each line of your audit, think about how time consuming each fix will be. You can use a rating scale where 1 is the least time and 5 is the most time. By sorting the ratings, you can find quick wins. If the attribute is a 1 or a 2, look at the other attributes to find where you can achieve quick wins. For example, if there is a page that has a 1 and is an important page with a high drop-off rate, you may be able make a small improvement that will lower your drop-off rates and improve your progress toward meeting business goals.

One trick I use when looking at audit results is to add the numbers both horizontally and vertically to give you sum of all the scores in each row and column. This can help you dig out which pages have the most issues. If you have pages with sums that average below 3, you'll definitely want to explore what's going on. Additionally, by looking across the bottom of the columns for each sum, the columns with the lowest sums point to potential problem areas throughout the site.

If you have a corporate website with hundreds of pages, a content audit be daunting. Where do you begin? What can you learn from this spreadsheet full of numbers? However, there are some simple ways to address what can seem like an overwhelming project. These tricks and tools of the trade are key to helping you get started with your analysis of the audit.

If you used a 1–5 scale, you can sort the spreadsheet by the column you want to analyze and see where all your 1's and 2's are under that attribute. For example, if you want to know more about where people drop off the site, you can sort the drop-off rate column and hone in on your lowest scores. This will tell you where people leave a page and help you hone in on any pain points. If you have high drop off rates on an important page, such as sign in or place order, you'll need to put that at the top of your to-do list.

Are there any patterns, such as clusters of pages where people are dropping off, user flows with low usability, areas of low findability? This may give you the opportunity to consolidate improvements to one specific part of the site at a time.

At TIAA, we were having drop off at various points during transactions. Once an audit honed in the areas of the user experience and critical pages that weren't working for our users, we got to work addressing those places. This all happened in the midst of a brand makeover for the website, so the timing for making changes was perfect. We could throw out anything that wasn't working and start fresh on new concepts.

Analyzing the results from your audit is gratifying work. Though it can take a lot of time, it's time well-spent. You'll learn a lot, not just about the quality of your site, but also about what makes for good content and what does not.

Although your recommendations ultimately affect individual pieces of content, consider what they tell you about the product as a whole. It's easy to get lost in the weeds (not to mention overwhelmed), but if you can start with that broader picture, those insights will generate the most meaningful ideas for long-term change to your content and processes.[1]

---

[1] *Content Audits and Inventories* (Land 2023, p.170)

The most important things to look for in any given audit are going to vary depending on the type of project you are planning and what questions you need to answer for that project. That might be related to a particular content type, a particular audience, the structure of the content, the style in which it's written, and so on. But one tactic that I always recommend is to look for patterns—if you can detect particular patterns that allow you to address content issues in multiples vs. single pages, you can save a lot of time in the next phase, which is when you begin to address the issues you find. For example, if you were able to identify that all content related to a particular product is outdated, you can chunk that work up and improve it as a whole.

An audit provides a baseline of the current state of content against which future content can be measured, so it forces you to define for your organization what your standards are and offers a framework for tracking content against them. The objectives you set for your content and the criteria you use to evaluate it ultimately help define how you want to position yourself as a business.

—Paula Ladenburg Land, Principal Consultant, Enterprise Knowledge, LLC

Make a plan

Now that you've sorted through the audit, how do you want to approach the work? You can scan the columns for the lowest rated pages, or the column that summarizes the ratings, and dig into how much work it would be to fix those pages and how critical those changes are for achieving business goals. This will help you develop a strategy for improving the website. Beyond quick wins, you may find that the pages rated as 3–5 in terms of difficulty to improve will require a deeper look. Ask yourself how much each of these pages disrupts the user and whether or not it affects other user flows.

Always keep the users front-of-mind and balance their needs with the business goals you gathered from stakeholders. Some items you may not be able to balance. Those you can take directly to stakeholders for discussion in your presentation meeting.

Present to stakeholders

You present your plan to stakeholders with the goal of reaching agreement on a final plan. The stakeholders will sign off on time and budget for some of your planned items and not others. They may find some improvements and fixes more or less critical based on their insights into the business, but hopefully everyone leaves the presentation with a deeper insight into the audit, the website, and the business in general.

> Please be sure to visit Chapter 8, *Learn who your users are*, for information on meeting logistics and the role of each stakeholder in the content design process. This will help you get the feedback and information you need to get the most out of your analysis.

Your presentation can have slides for each area in your agenda. Rather than show the entire audit spreadsheet during your presentation, you can make it available after the meeting and, instead, include snapshots that encapsulate the points you're trying to make. For instance, if each page in a flow has a low score and an easier fix score, then you may want to demonstrate that with a screenshot in your presentation. Or perhaps make screenshots of places in the audit where you found quick fixes that can make a big difference in achieving business goals. You can also make screenshots of the web pages that you are recommending be changed. This will help stakeholders visualize how the proposed changes will affect the user flow.

Stakeholders are looking out for business goals and the user (in that order), while content designers are looking out for the user and the business goals. What you can learn from your product and business stakeholders is insight into future plans, the most important areas to focus efforts on, and a sense of the amount of time and budget they're willing to commit to the plan.

Once you walk through your findings and the draft of your plan, it's time to discuss the plan with stakeholders and answer any questions they may have. Listen to what they have to say before responding. Weigh their insights into the business and the user against what you found in the audit. Through constructive conversation, you can arrive at a final plan and agree on next steps.

You can schedule a brief meeting to get stakeholder signoff on the first, easy-to-implement changes, then, after the changes are implemented and you have a set of metrics to compare to the audit, you can evaluate results. Based on your findings, you and your stakeholders can take another look at the plan and agree on how to proceed.[2]

---

[2] For more on presenting audit findings, check out Paula Land's book, *Content Audits and Inventories* (Land 2023).

TIDAL BANK

Case Study: Content audit

Once Anne and Santos have a solid content inventory, they start working on a content audit. First, they identify what attributes they want to use to evaluate the strengths and weaknesses of the content on the site. They come up with following 14 categories:

- Accessible
- Inclusive
- Usable
- Readable
- Findable
- Relevant
- Accurate
- Trustworthy
- Timely
- Clear
- Necessary
- Has appropriate tone
- On brand
- Provides business value

They add these attributes as columns to their content inventory, so they can keep all of the information in one place. They use a scale from 1 to 5 to evaluate each page and enter the ratings into the appropriate columns on the spreadsheet (see Table 7.1).

Table 7.1 – Tidal Bank content audit spreadsheet

URL	Page Title	Accessible	Inclusive	Usable	Readable	Relevant	Accurate	Timely	Clear	Necessary	Appropriate	On brand	Business value
/accounts	Accounts	2	1	2	4	2	5	3	3	1	2	2	2
/checking	Checking	3	2	3	2	3	2	3	2	3	2	1	2
/saving	Saving	3	3	2	2	2	3	3	2	3	2	2	2
/certificate-of-deposit	Certificate of Deposit	1	2	2	3	4	4	3	3	3	2	2	3
/money-market	Money Market	2	2	3	3	2	3	3	2	2	2	2	4
/what-are-cds	What are Certificates of Deposit	2	2	1	2	2	3	2	3	1	2	2	2
/CD-FAQs	CD FAQs	1	2	2	3	3	3	2	3	2	2	2	3
/What-account-is-right	Accounts	2	2	3	2	3	4	3	4	3	3	3	3
/contact-us	Contact Us	2	2	2	3	4	3	4	3	5	2	2	5
/rates	Rates	1	2	3	3	4	5	3	3	5	3	2	4

After reviewing all of the content for the audit, Anne and Santos have a clearer picture about what's happening on each page. They notice that the Account page rates very low on Usable and Necessary; this may be why it has such a high bounce rate despite being a highly visited page.

When they look more closely at the Certificates of Deposit pages, they notice that the main Certificates of Deposit page ranks high for being Relevant and Accurate and higher on most of the categories than the average Tidal Bank page. They think that one of the reasons this page rates so highly is because the CD rates are listed right at the top of the page (see Figure 7.1).

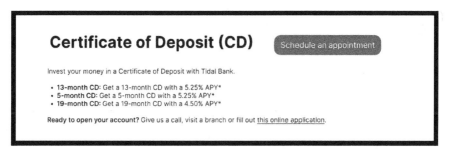

Figure 7.1 – Tidal Bank CD web page

They also notice that the CD FAQs page and the What is a CD? page rank very low in most categories. Anne and Santos consider that maybe these three pages could be combined to create a more useful experience for users.

After reviewing all of the content for the audit, Anne and Santos see several areas they can improve with good content design. Many of the pages on the site feel unnecessary and do not add business value. There is also a wide range of tones used across the site, which Anne wants to address. Plus many of the pages are not accessible or inclusive to all audiences.

Anne and Santos also see opportunities to reduce the number of pages and make sure all of the pages serve a clear purpose for users and provide a seamless experience for them to open an account.

CHAPTER 8

Learn who your users are

Understanding your users, who they are and what they want, as well as understanding your business goals will help you design best-in-class digital products. There are a lot of ways to do that. In this chapter, I cover the following, which I think are the most important:

- User journeys
- User interviews
- Competitive analysis
- Empathy mapping

User journeys

A user journey, or customer journey, is a map of the steps a user takes to achieve a particular task across various touchpoints within a company. The journey map tells the user story through each phase of their journey. The map doesn't just record each step a user takes; it helps you empathize with the user's thoughts, feelings, and actions. Everything you learned in the user interviews will come into play as you map out the user journey.

Here is an example of a user journey: a potential user receives a flyer from Tidal Bank with a high interest rate for new savings accounts. They decide to look into the bank online and follow the website to the account page to learn more about their high interest savings accounts. Once they're convinced that this is something that would benefit them, they look for an account opening page and begin the process, continuing until they either open the account or abandon the process.

Understanding user journeys is essential for any content-first design project because it is the bridge between the user's actual experience and all the different stakeholders, touchpoints, and concerns that support it. These include not only UI, channels, and content, but also key messages, calls to action, data generated—like metrics and analytics—and data required to adapt and improve the experience—like CRM data, social listening, entry URLs, content engagement history, and more. A user journey is the detailed and living documentation of the 5Ws of a user's experience achieving a key objective—the who, what, where, when, why of content.

Journeys are absolutely not sales funnels overlaid on an editorial calendar. It's important to remember that it's "user journeys," plural, not "the user journey." Users have many journeys over the lifecycle of their relationship with the brand. There are relationship-level journeys that are high-level and do resemble a funnel, with stages like "awareness," "consideration," etc. but the most important ones are the more granular journeys like "Choosing a mortgage", "Switching web services suppliers," "Choosing a major at university," and so on.

Finally, for detailed UX writing or feature- and section-level content, there are journeys that describe the experience of specific tasks like, "registering as a new user" or "upgrading an account."

—Noz Urbina, Founder of Urbina Consulting and co-author of *Content Strategy* (Bailie & Urbina 2012)

Well-defined user journeys set the stage for true content value design, because they support the creation of contextual content that speaks directly to the user's information needs at each stage of their interaction with a brand or product. they lay out user questions over time, creating a dialogue where content and UX answer user questions using all available context cues to deliver the right content, format, and next-best actions.

Looking at what questions certain personas have and how those questions might play out across time and channels can help you define the depth, breadth, and sequences that will drive the most valuable experience for user and brand.

For example, if the overall journey is choosing a mortgage, questions such as, "What kinds of mortgages are there?," "Am I eligible for a mortgage?," and "How to prepare for a mortgage discussion with my bank," indicate that the user wants to self-educate and would benefit from a beginner's tutorial and options that build trust and a brand relationship.

The questions that a user asks, as well as the rest of the 5Ws around that question, inform:

- **Empathy and accessibility:** through thorough exploration of users' experiences and needs
- **Current content analyses:** for audits and fit/gap
- **Content structures:** to support progressive disclosure, content personalization, and omnichannel experiences
- **Content labeling and language:** to understand different users' mental models and keywords
- **Content measurement opportunities:** to understand patterns and deep insights into user behaviors.

—Noz Urbina

User interviews

Effective user interviews provide a deep understanding of your users' thoughts, feelings, and actions. You can conduct them at any point in the design process, though they're most helpful in the discovery, design, and usability testing phases.

If you are interviewing users about an existing digital experience, explore what they expect, what they feel at each point in the flow, and what their pain points are.

If you're interviewing users for an upcoming project, lean more toward questions that explore expectations, hopes, discomforts, or fears. Find out what tasks they're hoping to accomplish in the experience you're designing, what they don't want, and what their ultimate goals are.

There are many reasons to conduct user interviews, some which are to learn more about:

- Users' pain points during an existing experience
- What was memorable about an experience and why
- What they like and don't like
- What they care about
- What are their motivations and concerns
- What were their expectations of the user interface and how did they respond to it
- What do they wish for in the digital experience
- Whether or not they find your digital experience effective
- User habits

You can also use these interviews to inform your:

- Problem statement
- Content audit
- Story mapping
- User stories
- User-needs statements
- Empathy maps
- Personas
- Work priorities and priority guide

How to conduct a user interview

Here are some guidelines for conducting these interviews. The most important task is to define a goal. What do you hope to get out of the interview? From there you can begin to outline your questions. Selecting participants will depend on who your target audience is. Typically a user research colleague will do the recruiting based on demographics as well as people interested in aspects of the experience you're designing.

Writing the questions

Stay focused on the user's perspective by asking neutral questions, not leading questions. You need their answers to be as unbiased as possible. Make the questions friendly and approachable, and don't place any pressure on the participant to answer every question or "get it right." There are no right or wrong answers. Keep your questions brief and focused on what you hope to learn from the interview.

Be organized and clear about what you hope to learn in the interview while also being open to going down one or two rabbit holes that your instinct tells you might lead somewhere helpful. Curiosity is the main emotion to lean into.

Also be aware that when asking questions that require the interviewee to imagine the future or remember the past, they will do so with an eye toward what they currently feel and skew the answer as a result. Someone asked to remember why they bought a particular car will invent reasons that may not have been apparent to them at the time to justify the purchase. Best to also ask counterfactual questions like "Was there anything that made you almost *not* buy the car?" to force them to imagine alternatives to what they currently feel.

—David Dylan Thomas, Founder and CEO at David Dylan Thomas, LLC, author of *Design for Cognitive Bias* (Thomas 2020)

Conducting the interview

Be friendly and approachable when you introduce yourself, and let the participant know that there are no right or wrong answers. Let the participant lead the interview. The most important thing you can do is *listen*. Don't interrupt. This is harder than it sounds, but it is critical to the success of the interview.

You will find places in the participant's responses where you can ask follow-up questions. Look at your list of questions as a roadmap for the conversation. Understand that you may not get to everything because you've explored other relevant avenues of inquiry, but that's okay. You're gathering important information.

Analyzing the results

You may notice particular themes emerge once you step back and look at all of the interviews. Group themes by common keywords. Also, grab quotes that illustrate the problems you want to address in your design. Where does your product need improvement? What underlying assumptions did you make before talking to users, and did the interview confirm those assumptions?

As you conduct your analysis, keep in mind what you set out to learn. What are your findings? Are there user insights that you didn't expect? The result of your analysis will point out user pain points and areas for improvement and help you put together a list of what you need to do next.

TIDAL BANK

Case Study: User interviews

Anne works with Millie, the UX researcher on the team, to conduct interviews with current and potential Tidal Bank users to better understand their needs when it comes to the Tidal Bank website.

Millie and Anne ask about the types of financial products the users have or are interested in. They ask about what types of information the users want when making decisions about financial products. And they ask about the users' previous experiences with banking and financial service websites.

Then they watch as the users use the Tidal Bank website to try to open a Certificate of Deposit and talk through their experience.

Millie and Anne come away with several key takeaways from the user interviews. Several of the users mention not knowing the difference between a CD and a Money Market account or not knowing which one would be best for them. The team notes this as an area where they could better inform their users.

Another takeaway from the interviews is that users have difficulty understanding how to open an account. During the walkthrough, several users said they felt unsure about where to go next and spent time clicking around the site trying to find the right next step. One user even gave up trying to open an account.

Millie and Anne recount the feedback and takeaways to the entire team. Then, Anne uses all of the feedback and qualitative research they've gathered so far to start re-imagining the user experience for Tidal Bank.

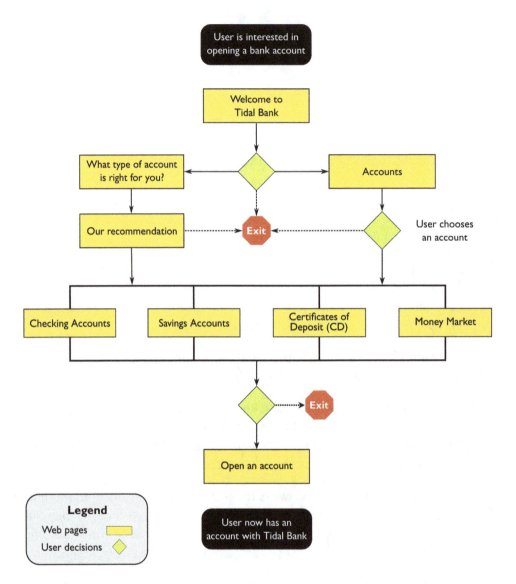

Figure 8.1 – Tidal Bank user journey to open an account

To do this, she builds a user journey for opening a new account (see Figure 8.1). By building out all of the steps in the process from the user perspective, Anne can explore what content users might need at each point in the journey.

Anne identifies several important pain points in the journey, including under-standing what account is right for the user and what information is needed to open an account. Once she has a user journey laid out, she shows it to the team for feedback and to ensure she has captured the full journey.

Once the user journey is finalized, Anne can use it to help design the pages and screens to make sure they meet user needs and answer the questions users may have throughout the process.

Competitive analysis

A competitive analysis evaluates similar products or experiences for strengths and weaknesses. It will help you gain insights into how other companies address their users' needs, preferences, and expectations as well as what does and doesn't work in their user experiences. This can provide valuable information as you move forward with your content and design decisions.

Before you get started with your analysis, outline your goals. What do you want to learn from this research? For example, do you want to understand more about why your customers drop off before choosing to open an account or place an order? Or perhaps you have more general questions about where user flows feel seamless and what elements or actions create friction. Friction can mean the difference between a sale and a frustrated potential customer. Friction can mean the user moves onto the next bank to open an account, but why? That's what you can learn through a competitive analysis.

When deciding what sites to analyze, look for businesses and products with similar activities and expected outcomes. They can be direct competitors or indirect competitors (for example, a company that offers a similar product, but in a different location). You should also consider what your site has in common with the competition and where is it different.

Here are some questions to ask when conducting the competitive analysis:

- What are the strengths and weaknesses of each competitor's product?
- How do competitors handle the pain points you're trying to solve?
- What are the characteristics of their content that work best and that don't work well?
- How does the tone of each competitor's content compare with your content? Is your content more or less on brand than theirs?
- Is there language for similar scenarios on your site that the competition handles better?
- What do they do well that you don't?
- What do you do well that competitors don't?
- Do you see anything that blows your mind? If so, break it down and ask why does it work so well? How can their success inform your content design?

The ultimate question is whether you can find things in the competition that can help you and your users in any way. Where can you find opportunities to gain an edge?

Once you've had the opportunity to analyze the competition, you can look for recurring themes and group them together. What can you learn there that will help you with your upcoming designs? Make a list of your findings, what to take with you and what to leave behind. What can give you an edge over the competition and what can simply help you tackle your problem statement more effectively and with fresh insight?

TIDAL BANK

Case Study: Competitive analysis

Now that Anne has a clearer picture of the Tidal Bank website and what possible improvements they might make, she conducts a competitive analysis to understand how Tidal Bank stacks up to competitors. She selects several different types of competitors, including direct as well as indirect competitors—competitors that serve slightly different products or to a slightly different audience.

The competitors she chooses are:

West Credit Union	West Credit Union is a regional credit union that offers checking and savings accounts, as well as other financial products. It is a direct competitor, serving the same community and offering similar products.
Seafarer's Bank	Seafarer's Bank is a local community bank that primarily offers checking and savings accounts. It is smaller than Tidal Bank or West Credit Union, but it is a direct competitor, serving the same community.
Web Bank	Web Bank is an online-only financial institution that offers checking and savings accounts. It is a national bank that has no offices in the community. However, it serves serves the same users as Tidal Bank and, therefore, is a direct competitor.
Big Corp Financial	Big Corp Financial is a large regional bank that offers checking and savings accounts, but also loans, credit cards and investment accounts. It offers similar products but is not available in the community that Tidal bank serves, so it is an indirect competitor.

Anne chooses these four competitors in order to better understand the things Tidal Bank's direct competitors are doing and to get ideas from indirect competitors about how Tidal Bank could improve. By doing a competitive analysis, Anne is able to see how Tidal Bank compares to other banks offering similar products. She's able to see what other banks are doing to answer similar problems and see where gaps might be across the entire industry.

She builds a spreadsheet to capture the following information:

- Institution name
- Institution URL
- Description
- Strengths
- Weaknesses
- Opportunities
- Threats
- What content works well
- What content doesn't work as well
- Takeaways

Competitive evaluation results

West Credit Union

Description: Small community credit union offering checking and savings accounts as well as loans and other financial products

Strengths: Clean design, bold simple language; starts main page with introductory information, which is customizable

Weaknesses: Pages are overloaded with information and the navigation can be somewhat circular at times

Opportunities: Application to open an account is in an entirely different flow. It is very jarring to move from the main website to the application

Threats: Outlines clear pros and cons and has a lot of detailed information about why to choose West

What content works well? Content is descriptive and is divided into chunks

What content doesn't work well? Too much information on a page, although it is chunked, it could still be a bit overwhelming. Descriptive content is separate from application process

Takeaways: Break content up but beware of still having too much information on a page; keep information relevant to the application close to the application

Seamen's Bank

Description: Small community bank offering checking and savings accounts

Strengths: Focused on community and trust in main content; uses bullets to outline different accounts in one place

Weaknesses: A lot of community content on the main page, not as much information about what types of accounts are available and how to open an account; doesn't allow you to start an application online from the accounts page

Opportunities: Impossible to start the application process online, you must go into a branch to start, but it doesn't tell you that clearly on the website

Threats: Feels community minded; clear about where your money is going

What content works well? Content that shows the value of keeping money in a local bank; information on accounts is chunked and simple

What content doesn't work well? too much on each page and next steps unclear

Takeaways: Show the value of the institution; why pick this bank; have clear Calls to Action (CTAs) and Next Best Actions (NBAs)

Web Bank

Description: Online-only bank offering checking and savings accounts

Strengths: Upfront about the products it offers and how they save the customer money. Lots of clear language around how much customers will pay and how long setup will take

Weaknesses: A lot of information on the accounts page, feels built around SEO rather than user experience

Opportunities: Clear CTAs but application is still one step away from information about account

Threats: Competitive products, clear CTAs at the top of the page; application includes relevant information about what you need to start the process

What content works well? Gets to important information quickly and is chunked in easy to read sections; clear CTAs and NBAs

What content doesn't work well? Feels impersonal, focused on features rather than why users might want to choose the product

Takeaways: Approach customers where they are; use language that users can relate to

Big Corp Financial

Description: Large, national bank offering a wide range of personal and business banking options, including checking accounts, savings accounts, credit cards, and loans

Strengths: CTA to get started is on the main page, plus a checklist to follow along with. Clear value propositions for users about why they should choose this bank

Weaknesses: Focused less on the financial products and more on the technology of online banking; checklist is a PDF rather than an interactive web page

Opportunities: Account applications must be made at a branch office; every link opens a new window

Threats: Online banking setup is simple, has clear steps and assistive information as well as a progress bar to show how far through the process you are. However, you have to go to a branch to get started

What content works well? Content answers a lot of questions up front and has clear value props, CTAs, and NBAs

What content doesn't work well? Navigation feels bloated and there is probably more information on the site than absolutely necessary

Takeaways: Beware of having too much content, but try to answer user questions before they are asked

Through this process, Anne identifies several recommendations for how Tidal Bank can enhance its user experience to better compete with other banks. For example, she notices that many pages on West Credit Union's website are very text heavy and include a lot of information. In some cases this actually makes it harder to complete tasks. Her biggest takeaway from West Credit Union's site is to keep information relevant to the application close to the application.

When she reviews Web Bank's website she notices how approachable the content is. As a web-first company, their website is the only place they can make an impression on their customers and potential customers, and it shows. They keep things clean and simple throughout. Anne's main takeaway from Web Bank is to make sure to approach customers where they are and to stay away from jargon, instead using language that users can relate to and understand quickly.

The competitive analysis gives Anne has several ideas she can integrate into the new website that will help Tidal Bank get ahead of competitors. Now that the team has clear data and analysis on the Tidal Bank website and its competitors, it's time to talk to actual users about what their wants and needs are.

Empathy mapping

An empathy map is a valuable tool to help you get inside a user's experience. Empathy mapping is a collaborative exercise conducted with users or the stakeholder team to gather information about how they think and feel and what they say and do. This exercise gives you a view into users' needs and their decision making processes, so you can make informed design decisions.

Empathy mapping requires you to have a user flow or user story that you want participants to carry out. As they work through the experience, have them narrate what they're thinking and how they're feeling as you record what they do and say. You can draw an empathy map on a whiteboard (see Figure 8.2), write responses on post-it notes, and place them in the appropriate quadrant. You and the participant collaborate on filling out the map, which helps ensure you agree on the completed map.

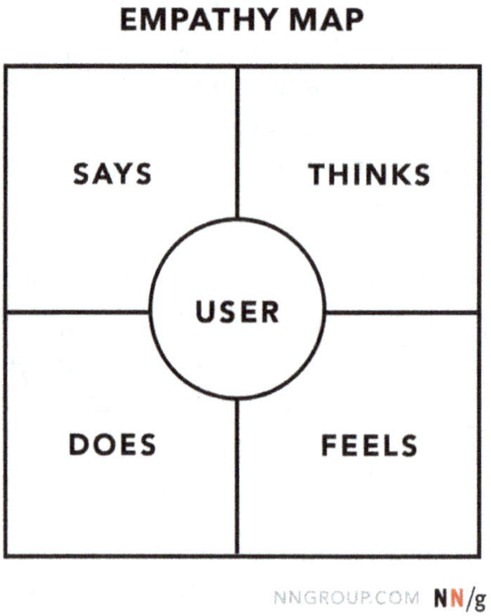

EMPATHY MAP

SAYS	THINKS
DOES	**FEELS**

USER

NNGROUP.COM **NN/g**

Figure 8.2 – Empathy Map[1]

---

[1] Image from "Empathy Mapping: The First Step in Design Thinking" (Gibbons 2018).

TIDAL BANK

Case Study: Empathy mapping

Now that the team has completed the user interviews and created a user journey, they decide to conduct an empathy mapping exercise. Anne and Millie gather five participants that fit the team's primary customers: **adults in their 20s who are just starting to think about their financial needs.**

Anne and Millie walk the participants through the user journey. As they walk through the journey, Millie asks the participants about the journey and asks them to narrate what they are thinking and feeling through the process.

As Millie leads the participants through the exercise, Anne writes the responses on sticky notes and places them in the appropriate quadrant on the empathy map. She uses a different color for each participant (see Figure 8.3).

After the exercise, Millie and Anne review the results. A few things pop out to them. They notice that the majority of their participants had some understanding of banking products but felt like they could learn more. They also notice that several participants said they would spend a lot of time reviewing products before opening an account and that the user journey allowed them to do that.

Overall, they feel that the feedback they received during this exercise confirms their direction in creating a user journey that highlights learning about different products. They take these results back to the team to review.

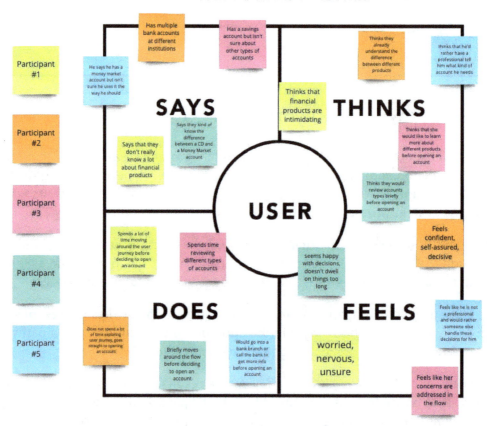

Figure 8.3 – Tidal Bank empathy map

CHAPTER 9

The design phase

Now that you've completed the research part of the double diamond with a define phase and a discovery phase, you're ready to get to work building a prototype with your UX collaborators. The first step is to create a priority guide that will lead to a rapid prototype. With a quick prototype in hand, the next step is A/B testing, which leads to successive prototype updates until you reach a solution that best meets the users' needs and business goals. Iteration is critical to this phase of the work because each iteration brings you closer to a solution.

Priority guide

A priority guide is a content outline of a page designed for mobile-first. It shows the content elements of a web page in priority order from top to bottom (most important content at the top). It does not dictate the visual design. Instead, it is meant to give web site designers guidance on the content, but leave them the freedom to create a compelling visual design.

When you create a priority guide, you take into consideration the type of each element on a page, the content that fills that element, and the hierarchy of elements on the page. This is all done using a content-first approach based on user needs. The Tidal Bank case study later in this chapter contains an example of part of their priority guide.

 Drew Clemens, who first created and coined the term *priority guide*, said: "Essentially, with the priority guide, we create a single deliverable that provides direction for content-focused design and mobile-first development in something resembling a wireframe."

—Drew Clemens, "Design Process In The Responsive Age" (Clemens 2012)

Drew designed the priority guide to use in a mobile-first approach because forcing a single column layout makes content hierarchy easier to work with, understand, and design from. The content hierarchy you develop will be based on user needs and the results of the research conducted in the define and discovery phases of the double diamond approach. In other words, here's where the rubber meets the road.

As you start the priority guide, continue to follow best practices for UX, UI, accessibility, and inclusivity to make sure you deliver a meaningful experience. Start by considering the goal of each page and ensure that every piece of content has a reason for being there based on solid research. Also consider functional content, such as dropdowns, links, checkboxes, radio buttons, and so on in the same way.

If you're struggling to get started, try writing a content outline. Visualizing a content hierarchy in outline form may help you move onto the priority guide. Regardless of how you get started, the priority guide is an important tool to use before you begin to sketch prototypes.

Presenting the priority guide to your team of stakeholders will help further refine both the elements of content and their hierarchy. Many stakeholders may not have worked with a priority guide before, so give them a brief introduction to the purpose and benefits of a priority guide to help them understand why their perspective matters.

What are the benefits of a priority guide? Priority guides provide:

- A straightforward format you can use to quickly discover and define the hierarchy of a site
- A solid starting point for creating a page's wireframes based on user research, business requirements, and consensus from stakeholders
- A page-by-page analysis that will help you design a prototype faster than if you start prototyping in the dark, and anything that saves time and budget will make your stakeholders happy

TIDAL BANK

Case Study: Priority guide

Anne uses the information gathered in the user journey to identify five main pages that she wants to focus on for a content redesign: the home screen, the Accounts page, the Money Market page, the Certificates of Deposit page and the Open an Account guide page.

Before starting to design these pages, Anne outlines the content that she thinks should be on each page. She builds a priority guide that shows the information structure for each page and the hierarchy of information on the page (see Figure 9.1 and Figure 9.2).

By building priority guides for these pages, Anne can help ensure that the designs will include the right information in the right order to meet user needs.

Welcome

- Navigation
- Login to account
- Tidal bank intro
- What do you want to do
- Types of services available
- How to contact us

Figure 9.1 – Priority guide: Home page

> Heading
>
> - Navigation
> - What does it mean to open an account?
> - What types of accounts does Tidal Bank offer?
>   - List of accounts with brief descriptions
>   - Links to in-depth account pages
> - Ready to open an account?
> - Not sure what is right for you? Take our quiz
> - How to contact us

Figure 9.2 – Priority guide: Accounts page

Rapid prototyping

Rapid prototyping can take place anywhere, on anything. It can take place in an airport bar on a cocktail napkin, on the back page of the novel you're reading on your train ride home from work, or at your desk in an application such as Figma. What's important is that you create, test, and iterate in rapid sprints. This means days rather than weeks, and as you near your solution, the sprints can be set to run for hours. Rapid means rapid.

As you sketch each page, keep in mind the goal of the page and the content hierarchy that you established with the priority guide. Remember, you want to start with the content—every digital experience is a conversation with the user, and conversations begin with words. Here are some questions to ask yourself before you start to sketch:

- What does the customer want?
- What are the steps to help them get what they want?
- What have you learned about how they communicate digitally?
- What are the steps to help them get there?

- What are the possible pain points?
- How do they know they are succeeding?
- How do you keep them engaged?
- How can this content help you achieve your business goals?

TIDAL BANK

Case Study: Prototyping

Now that the team has a clear understanding of what pages they want to focus their redesign on and the information and structure of those pages, they begin designing a first draft prototype in Figma.[1]

Santos starts building page designs based on the new Tidal Bank look and feel. Anne begins adding content to the pages based on the information hierarchy the group established in the priority guides.

There are several considerations Anne makes as she writes the first draft of content for the new pages. In addition to using the information hierarchy in the priority guides, she thinks through word choices that match the tone and voice of Tidal Bank. She also focuses on making sure the content is clear and that the information and actions are focused on user needs.

Figures 9.3, 9.4, 9.5, and 9.6 show the first prototype that Anne created.

Anne goes through several iterations in order to get wording right and to edit down content that is confusing or superfluous.

---

[1] https://figma.com

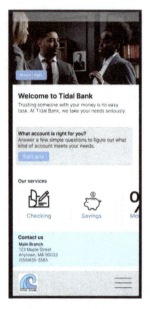

Figure 9.3 – Tidal Bank mobile prototype: part 1

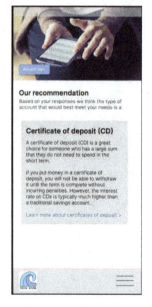

Figure 9.4 – Tidal Bank mobile prototype: part 2

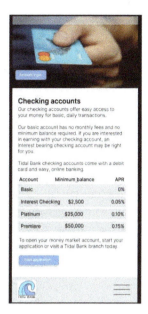

Figure 9.5 – Tidal Bank mobile prototype: part 3

Figure 9.6 – Tidal Bank mobile prototype: part 4

A/B testing

Once you have a couple of versions of your quick prototype, run some simple A/B testing with anyone at hand. You may use people at work, online users, or folks at home. Whoever you choose will provide valuable information about the direction your initial prototype needs to take.

A/B testing

Also referred to as *split testing*, A/B testing is a test in which you run two versions of a site or page with different users and collect and analyze their feedback. After analyzing their feedback you can consider which version performs better, version A or version B. While you can use more than one variable in your test, analyzing one thing at a time can give you more specific, directed feedback.

First, you need to define what "better" means. Here are a few things you might test:

- Which button content gets more people to open an account?
- Which headline gets more people to read an article?
- Which email subject line gets more people to open an email?

Keep the two versions as similar as possible, with the only variable being the content you want to test. Once you have the results of the A/B testing synthesized, you can return to your stakeholder group, present your findings, make suggestions, and ask for feedback. With the depth of user research you've done, and the testing results, you should be able to reach consensus about next steps and move on to your next prototyping exercise.

Prototype

Prototypes are a picture worth a thousand words. Simply put, a prototype is a simplified version of a product design, in this case, driven by content. The purpose of the prototype is to test the content and design with users and stakeholders, gather feedback, and hone in on the best solution. Your prototype will go through many iterations, each of which should move you closer to a solution that the team can sign off on.

> Prototypes slow us down to speed us up. By taking the time to design a prototype for our ideas, we avoid costly mistakes such as becoming too complex too early and sticking with a weak idea for too long.
>
> —Tim Brown, CEO and president of IDEO, author of *Change by Design* (Brown & Katz 2019)

Typically, designers participate in all of the research work, so they won't be surprised to receive your sketches. Many will welcome them as a great starting point as you work together toward final designs and content. If your designers don't welcome your efforts, point them to the research, show them your sketches, and share the results of your testing, then back up your ideas with facts.

When you and your designers are ready to commit the sketches and rough designs to a prototype, you can use a rough click through of a user experience, or a more refined click through using real design elements. The prototype gives you the opportunity to work through problems in your initial vision and gain further insight into solving the problem statement.

Iterate

You test the prototype and iterate the same way you test your sketches. You cycle through the process of prototyping, testing, reviewing with stakeholders, and prototyping and repeat until you reach a solution that the team will sign off on. You may need to repeat discovery or definition work, such as refining the problem statement, further exploring user interviews, and so on. The point is to keep repeating the design process until you reach a best-in-class solution.

TIDAL BANK

Case Study: Share the prototype

Once the team has a first draft of the prototype, complete with draft content, Santos and Anne share their work with the larger design team to discuss their design decisions and to get feedback and fresh perspectives.

Santos and Anne also share the design file, which they created using Figma, with the group so that others can review the designs more closely, add comments, and share other feedback. This feedback can include thoughts on how the page is laid out, whether design elements work in context, and whether the content meets expectations.

Melissa, the content design manager, reviews the Tidal Bank prototype and makes suggestions on how to improve the content on the page. She recommends changes that align the content more closely with the Tidal Bank style, notes ways to add clarity, and suggests places where the content could be simplified.

For example, the Accounts page is a bit content heavy and could be overwhelming for users. Melissa recommends cutting some of the content on the page to make it easier for users to skim and find the information they are looking for.

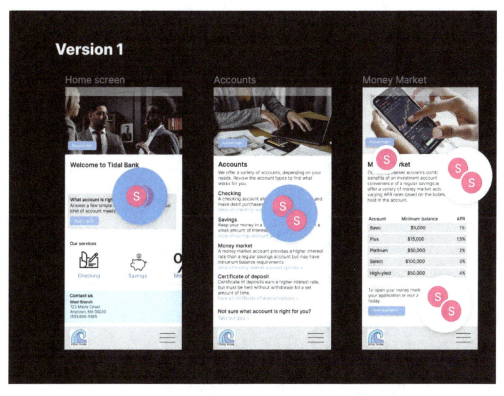

Figure 9.7 – Figma workspace showing the first version of the prototype

The team uses the commenting tool in Figma to highlight areas that could be improved, and they discuss what changes to make (see Figure 9.7). This makes it easier for the team to see exactly what each piece of feedback refers to and where to make changes.

For example, Figure 9.8 shows a pop-up screen with a comment from a reviewer. Other reviewers and the originator of the design can see each other's comments and reply to them.

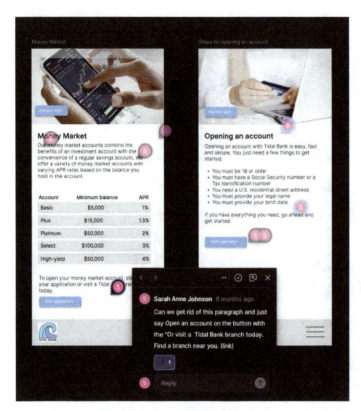

Figure 9.8 – Figma workspace showing a comment from a reviewer

The team takes all of the feedback from the design critiques, as well as all of the comments and considers how they might incorporate them into their designs. They make a new version of the prototype with these changes. Anne updates the Accounts page based on the feedback she received. She simplifies the content on the page to make it easier for users to understand (see Figure 9.9).

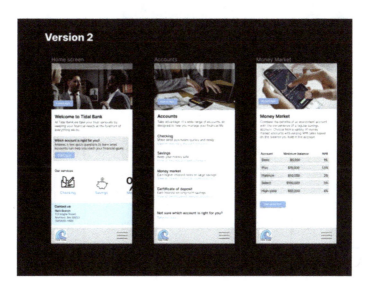

Figure 9.9 – Figma workspace showing the second version of the prototype

After the team makes all of the updates, they review the pages with the larger design team again to make sure they've addressed feedback and to understand if there are any other questions or concerns from the team.

Figures 9.10, 9.11, 9.12, and 9.13 show the final prototype of the Tidal Bank mobile website.

Figure 9.10 – Tidal Bank final prototype: part 1

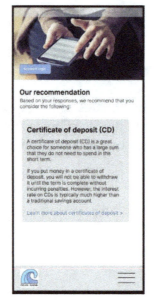

Figure 9.11 – Tidal Bank final prototype: part 2

Certificates of deposit (CD)

Certificates of deposit at Tidal Bank offer competitive rates for a variety of terms. All of our CDs are federally insured.

Choose a timeline that's right for you and your goals to lock in a guaranteed return.

CD term	Minimum balance	APR
13-Month Promo	$500	5%
3-Month	$500	2.3%
6-Month	$1,000	2.5%
Jumbo 12-Month	$100,000	2.75%
Jumbo 24-Month	$100,000	4%

Checking accounts

Checking accounts offer easy access to your money for basic, daily transactions.

A basic account has no monthly fees and no required minimum balance. If you're interested in earning with your checking account, an interest bearing checking account may be right for you.

Tidal Bank checking accounts come with a debit card and easy, online banking.

Account	Minimum balance	APR
Basic		0%
Interest Checking	$2,500	0.05%
Platinum	$25,000	0.10%
Premiere	$50,000	0.15%

Savings accounts

A simple savings account may be just what you need, whether you are saving up for college, a new car, or even a vacation.

Savings accounts are easy to open and maintain. Choose from a variety of accounts with different interest rates and minimum balance requirements so you can save your way.

Account	Minimum balance	APR
Regular savings	$500	0.05%
Platinum savings	$1,000	0.08%
College savings	$500	0.10%

Figure 9.12 – Tidal Bank final prototype: part 3

Contact us

We love hearing from and seeing our customers. Stop by our branches during regular hours or give us a call whenever you have questions or concerns. We are here to help.

Main Branch
123 Maple Street
Anytown, MA 00033
(555)655-5565
Monday-Friday 8:30 AM-5:30 PM; Saturday 9 AM-1 PM

East Branch
355 East Drive
Anytown, MA 00033
(555)655-6755
Monday-Friday 8:30 AM-5:30 PM

West Branch
222 West Blvd
Anytown, MA 00033
(555)655-5775
Monday-Friday 8:00 AM-4:30 PM; Saturday 9 AM-2:30 PM

Or reach us any time by emailing:
email@tidalbank.com

Figure 9.13 – Tidal Bank final prototype: part 4

CHAPTER 10

The delivery phase

Collaboration between content designers and development is crucial for delivering the user-centric and seamless experiences you've created throughout this process. The development team brings your designs to life through code. That's why you need to get them involved early in the process.

Part of working with development includes making sure you understand what they do in the design and delivery process and vice versa. This relationship helps content designers understand development constraints and helps developers gain more insight into design and user experience considerations.

To build a great relationship, take the time to understand the basics of what the developers do. Learn some of their terminology, such as *front end* and *back end*. They are focused on a different aspect of the project, which means you're likely to have miscommunications as you speak about the project. This is why it's so important to create a good rapport and open communication to facilitate collaboration.

Be sure that you agree on how you'll deliver content to the developers. Will it be in a content management system (CMS), a spreadsheet, a Word document, Airtable, or possibly Figma? Are you on the same page as to the best format for the documentation to be presented in? Developers are your gateway to the public. What they publish is what your users will see. After all of the hard work you've done to create content, you want to work closely with development to ensure that it's delivered correctly.

As you work through the design phase of the project, stay in touch with your CMS gurus and developers to figure out if the project will be implemented in phases for your review or all at once at the end of the design phase. Meeting with CMS developers and software engineers during the design phase can help ensure that the content and designs can be built using the tools and code available to them. You may want to include them in stakeholder meetings, especially design reviews, in which both content and design are presented. Finding a solution may require some back and forth, but if you work this out as you go, you'll save time during delivery.

Align with developers from the beginning to understand technical specs and any technical limitations. This ensures a shared understanding of the intent, scope and the desired outcome of the project, including how the developers will test and deploy the code.

Avoid making assumptions about the impact of design changes—what may seem straightforward and easy for you could actually be a much heavier lift for developers. If you're unsure, reach out and ask; it's wise to make sure developers have access to and can give feedback on early prototypes and deliverables.

—Aishling Seder, Aetna

TIDAL BANK

Case Study: Compliance and legal

They schedule a meeting with the compliance and legal teams to review the designs and make sure they meet compliance and legal guidelines. Because Tidal Bank is in financial services, there are legal restrictions and requirements for how the company presents certain information.

The legal and compliance teams review the new designs with the design team and make a few small recommendations for language changes to ensure that the pages are in compliance with the law.

Once the team has stakeholder approval and confirms that the designs meet all legal requirements, the team starts preparing the design file for hand-off.

Hand-off can mean different things to different teams, but for the Tidal Bank team, it means making sure everything is organized clearly in the design file, all of the content is up-to-date, and annotations are added to make sure the developers understand the intention of all parts of the design. This step includes things such as how a certain section might function or where a particular button leads users.

Once the design file is ready, the team schedules a meeting with the development team to walk through the design. This is a space for the development team to ask questions and better understand how to build the designs.

After the hand-off meeting, the development team begins building the new pages. As the pages are being built, the development team may share their work with the design team to ensure they are building the design correctly and that they are following the designs as outlined. This is an important part of quality assurance. This also provides an opportunity for content designer to copy edit the developer's work to ensure it is being implemented correctly.

Quality testing the content

Whether you have a quality assurance (QA) team or you're on your own, this phase is a crucial part of ensuring business success. Users make snap judgments in an online environment, and oftentimes will judge a site by the quality of its content. Quality creates trust and loyalty and brings people back to the site. That's what you want your product to achieve—return visitors.

Once a project is developed, it typically goes into an alpha, then a beta state in which each phase is run through a QA team. Your job then becomes fixing errors in the design or implementation of your product.

Sometimes, it's up to you to do your own quality assurance before and after your product launches. When you quality test your content, look for obvious glitches such as spelling and grammar issues, tone and brand adherence, misplacement or misalignments of words, and word length.

You need a process for testing the content. For example, you can run through each flow and make sure that what's on the screen matches what you wrote and that your content works well in the context of the design. Sort out with the development team how you will provide feedback and changes. You author any changes to your content, but you then need to communicate your changes to development, so they can make them on the production site.

Usability testing

I covered usability testing above, but now is the time to really dig in with real users to get the valuable feedback that will help you shape the content and deliver a product that helps users succeed in their tasks.

Metrics

Compare metrics from before delivery to those after delivery. Look for trends, improvements, and places that weren't as successful. Metrics can steer you toward areas that need attention, either from you or from other stakeholders.

A/B testing

Continue testing the changes you made in response to usability and metrics findings. The same A/B testing you used for prototypes works here, too. Simply run users through one version of the experience (A), then run them through the other contender (B) and compare findings. You may find that one clearly performs better than the other, or you may need to combine findings from each version and head back to the drawing board.

Iterate based on research and testing

As your testing identifies pain points and areas that need to be addressed, you can go back as far in the double-diamond process as you need. Some projects may require super quick turnaround, so you won't be able to go back as far as you might like, but other projects may have the time and resources to revisit prior assumptions.

You may need to modify the design to work on content changes or you may need to use some of the research techniques described in the define section. If you're really stuck or find that you're far off base, you may need to go back to the very beginning and re-articulate your problem statement. Whatever needs to be done, plan for it, allocate the time and resources, and dig in.

Any iteration is a step toward improvement. Your process may not be perfect, but you can make it work given your time and budget constraints, so go for it!

CHAPTER 11

Test and measure content at every phase

Test and measure content to ensure that the product answers the user's needs, to identify pain points, and to continue to be on the lookout for errors.

Page analytics can point to issues with content. For example, they can tell if users are reading an area of content, staying on the page, or selecting a button to continue in a task. You can also isolate content changes and run testing on only those pages with changes, which lets you gather comparative information you can use to elevate the user experience.

By mapping user questions to content as well as calls to action and data, teams can begin to work out what questions or actions indicate a user is advancing along their journey. One of the perennial challenges of content teams is measuring the success and value of content. The organization wants to know if relationships are developing and progress is being facilitated. Watching for journey progress signals can put numbers on content's impact.

Take for example, a mortgage page. If a user searches for rates at a specific bank, or hits a bank digital touchpoint and goes straight to comparison pages for specific mortgages, they may be ready for more advanced calculation tools or calls to action to contact an agent or broker.

This offers many metric options: how many visits does it take before users advance from self-education to being ready to engage further? What is the correlation between content engagement and journey progress or speed? What calls to action move people along the fastest?

—Noz Urbina, author, Founder Urbina Consulting

You can run tests with users by, for example, walking users through an experience and having them narrate how they understand the content. Do they understand what you're saying and what to do next? In this chapter, I also discuss a variety of other tests, including highlighter tests, recall tests, and cloze tests, also referred to as Mad Libs.

Testing, measuring, and gathering metrics give you the tools you need to advocate for content-first design in the UX organization and beyond.

Evaluating content

First, you need to decide what content you want to test and what you want to learn from each test. I start by evaluating the content using the team's guiding principles:

- Accessible
- Inclusive
- Usable
- Readable
- Findable
- Relevant
- Accurate
- Trustworthy
- Timely
- Clear
- Necessary
- Appropriate tone
- On brand
- Business value

When deciding what and how to test and measure, having a tightly scoped focus is key. You should know what you want to learn, but probably more importantly, what you don't want to focus on or can't focus on.

I've coached hundreds of content designers on how to do content testing, and there's a real temptation to test everything—all the words and terms that are important to your customer experience.

That's a recipe for burnout.

The good news is that the insights you uncover with content testing create so much energy and attention. The thing to be careful about is, that energy means you're going to be hearing from product managers and product designers and maybe teams you have never worked with before and even your executive leadership who will be asking for more content testing. You need to set expectations for them (and yourself of course), so be careful to bite off only what you can chew.

—Erica Jorgensen, author of *Strategic Content Design* (Jorgensen 2023) and staff content designer at Chewy

Highlighter test

In this content testing method you ask users to mark what's clear and unclear in a piece of content using highlighters; for example, green for clear and orange for unclear. This test helps you:

- Identify missing information that would help users complete a task
- Locate content that needs to be reworked
- Simplify text to provide only what's essential to users

Here's how to run a highlighter test:

- Decide what content to test
- Set the user up with instructions and highlighters
- Before they highlight the content, have them read it through
- Have them highlight the content
- Ask and answer follow-up questions
- Gather qualitative data to review with stakeholders alongside the highlighted text

Cloze test

This is also referred to as Mad Libs because it requires that you take a body of text and remove words for the user to fill in. This test measures users' understanding of the context and meaning of what they've read. These are great tests for larger pieces of content or complex content that can be difficult to understand and parse.

Here's how to run a cloze test:

- Choose the content you want to test
- Delete every fifth or sixth word so you have plenty of blanks
- Have the user fill in the blanks to the best of their ability
- Score the test

According to Nikki Anderson, you can score your test like this:

- Count the number of correct answers and divide that by the total number of blank words
- Turn this number into a percentage. For example, if there were 25 blanks and a participant got 15 right, they would get a score of 60%.

After scoring, you can use the following benchmarks:

- 60% or higher indicates the text makes sense to the audience
- 40% to 60% indicates readers might have a hard time understanding the text
- Under 40% indicates people will struggle understanding the text

Recall test

Recall tests are just what they sound like. You have users read through a body of text and ask them if they understood it and what it means to them.

Here's how to run a recall test:

- Use content that has previously been a pain point or new content you want to test
- Have your goals set. What do you want them to read and understand?
- When the reader is done reading the content, ask fact-based questions.

For example, at TIAA, at the end of the Required Minimum Distribution content work, we asked users what they just read and if they understood it. This pointed us to pain points, gaps, and places where we needed to work on the language. The test proved valuable in making content improvements.

Other ways to test your work include content heuristics, content scorecards, design critiques, and peer-to-peer reviews. These evaluations also help stakeholders understand why you've made the content decisions you have, and they provide context like metrics and content audits do.

Content heuristics

Content heuristics create alignment around content best practices in your organization. They create consistency of voice, tone, and intention. Your list of heuristics does not have to be extensive, but it should capture the essential elements that make for excellent content in your organization.

Heuristics

Heuristics provide guidelines that create consistency across all content and help you adhere to tone and voice standards. They are short cuts or rules of thumb that can give you a pragmatic evaluation of a piece of content.

Table 11.1 contains some heuristics you can use to evaluate content. These are just a few of the heuristics you can use for evaluation. You can also use the tests for evaluating content discussed in the section titled "Evaluating content" (p. 110).

Table 11.1 – Heuristics for testing content quality

Heuristic	Definition
User-centered	Keeps in mind user thoughts and feelings, readability level, and comprehension level. Understands what the user is trying to accomplish and makes it as plain and direct as possible for them to accomplish a task.
Simple and clear	Written in plain language that doesn't use jargon, acronyms, or language above the reading level you have defined for your audience.
Human	Is conversational and written in plain language with a focus on building user confidence.
Visual hierarchy	Uses clear visual cues that lead users to the actions they need to take to complete a task or find information.
Consistent	Uses the same words to express the same things; for example, labels buttons that do the same thing with the same words.
Accessible and inclusive	Doesn't use idioms or colloquialisms; uses familiar language that anyone can understand.
Readable	Can be easily read by your target audience. According to nngroup.com, "People understand 8th-grade–level content equally well on computers and mobile devices." However, be aware that your audience may be at a different level.
Empathetic	Keeps in mind what your users are trying to accomplish and what their pain points might be. How do your users feel when spending money on your product. How do they feel and think moving through your user flow?
Actionable	Leads users through a task or provides information for a reason. Next steps should be clear and make sense to your audience.
Scannable	Breaks up the cognitive load for users by helping them scan for what they need.

Content scorecards

Content scorecards rate content against a set of content principles, typically some or all of the heuristics. Choose a scale (e.g., 1–5 or 1–10) that is comfortable and practical for the amount and type of content and the heuristics you're evaluating. Table 11.2 shows a sample scorecard.

Table 11.2 – Sample content scorecard

Heuristic	Rating	Notes for improvement
User-centered		
Simple and clear		
Human		
Visual hierarchy		
Consistent		
Accessible and inclusive		
Helpful		
Empathetic		
Actionable		
Scannable		

Design critiques

Design critiques evaluate a design from an outside perspective. You bring together a group of content developers who have not worked on the design being critiqued. A representative of the team that created the design presents the critique team with the problem statement, then walks them through the define and design phase.

Establish ground rules before the meeting to create a productive and collaborative atmosphere. I ask participants to listen closely to the presenter's explanation of their design decisions and to what they're looking for in terms of feedback. Ask the critique team to provide on-topic feedback that begins with saying something positive before delivering productive commentary.

You can use the content scorecard to have people rate the content against heuristics or run a workshop where teammates share work and ideas. Both of these can improve work and initiate insightful conversations in which everyone learns something new.

In my managerial role, I called these sessions *design collaborations* because the term sets a more collegial tone that *design critique* and because the team began to collaborate to solve whatever wasn't working in their colleague's content. These sessions were fun and non-intimidating, and the outcomes were fantastic. Even if the person showing the work had to go back to the drawing board and work with the team's ideas, they gained new perspectives and ideas.

Peer-to-peer reviews

Peer-to-peer reviews usually happen among team members. Peer-to-peer reviewers can use the content scorecard to measure against heuristics, or they can provide direct feedback and work with teammates to improve the content. This method works well with teams that are looking for one-on-one or small-group collaboration.

Whatever evaluative methodologies you put into place, you'll gain insight that will improve your work. Use these methods at any or every phase of your content development, and you'll be surprised by what you learn.

 I don't consider content to be a separate part of the experience, so when I'm looking at the efficacy of my team's content I'm looking at the success of the experience as a whole. So my questions, if we're not meeting targets, will be around what in the experience, holistically, is causing drop-off or confusion? Do we need to re-examine placement of our entry points or is there an opaque UI pattern? Of course, we can and do iterate on content and narrative flow to improve things like conversion or engagement, but the way my team works is so embedded with our design partners that we measure the success of the overall experience.

—Selene De La Cruz, Sr. Manager of Content Design at Robinhood

I have used data points from whole pages to analyze the content on the page. For example, at TIAA when we were redoing the entire site, we were able to hone in on content that could be a challenge to the user. Also, you can use qualitative data such as scorecards and usability research, discussed in the section titled "Qualitative research" (p. 120).

Quantitative research

Quantitative research looks at numerical data to provide an objective perspective on user behavior. You can ask questions such as:

- How long did users stay on the page?
- Did they take the path you expected or intended to complete the task?
- How long did it take them to complete the task?
- Where were the pain points?
- Could users find the page they needed to complete the task?

These metrics provide numerical evidence that can point to both content and design areas that need to be addressed. It's important as a content designer to look at metrics and analyze where they point to content might not live up to your content standards. For example, if a button will take users to the next step in a flow but they drop off at that point without selecting the button, could it be due to the content?

Once you've evaluated your content against the data, you can work with your UX partners to find effective solutions to improve the user experience at trouble spots. Your product and business teams also have goals for the pages and the task(s) at hand. They will compare quantitative results against their Key Performance Indicators (KPIs) to identify places where they want the team to focus their efforts.

Your team can also use quantitative data to test against competitors' products that are trying to achieve similar goals. Measure your metrics against the metrics you find on a competitor's site and identify places where you're coming in low. This type of competitive research can provide insight into how other companies or organizations deal with the pain points you've discovered through your research.

It's important to weigh the pros and cons of conducting quantitative research before you dive in. Look at the goals of your research and what you want to determine, then make a decision as to whether this kind of research will help answer your questions.

Benefits of quantitative research

- Numbers and real data can help convince stakeholders about wins and losses of your design so that they can decide where to invest
- Avoids biases and human error by providing objective data
- Data is easy to analyze in charts, graphs, and spreadsheets
- Provides a way for you to evaluate your updates against your previous design

Potential pitfalls

- Requires a large sample of users, which can be expensive and take up time to manage
- Doesn't provide contact with users to ask questions that could give you insight into their behavior or to provide quotes to help stakeholders understand results
- While quantitative data provides results that you can use to identify potential content issues, it doesn't explain the patterns in the data or tell you how to address issues. You have to rely on your own expertise and insights

When to use quantitative research

Decide when to use quantitative research based on the pros and cons weighed against the goals of your project. Consider conducting quantitative research when you need to:

- Evaluate whether a redesign is necessary based on objective data
- Compare design solutions as you address findings from data and measure whether you've made an impact or not based on the numbers
- Prove the value of content by making content changes and re-conducting the research
- Compare your solution to competitors

How to measure content often boils down to: what tools do you have available? What access do you have to customers (or people who are very similar to your customers)? UX teams in regulated industries like finance and healthcare need to be especially careful when sourcing participants for usability testing.

Then there's a question of budget; a lot of teams are seeing cutbacks in budgets for tools like dscout and UserZoom, which is a shame. If you are fortunate to have access to platforms like those, great. If you don't, then make friends with your customer service team and see what data they have. They're often a goldmine of both quantitative and qualitative data and feedback from customers, so it's often smart to start there and see what information is immediately available to you, especially information about what specific content is leading to customer confusion.

Often what customer service data reveals is gaps in content—for example, a question that hundreds or thousands of customers are asking and that's not yet addressed by the info on your app or website. If you and your team can address those gaps, you'll see an immediate drop in customer-service contact volume for that particular topic, and you can measure the impact of your work.

For example: reducing calls on Topic XYZ by a volume of 1,000 calls a month, multiplied by the cost of a typical customer service call or chat will give you the cost savings you just created for your company. If a customer service call or chat costs $40, you've saved your company $40,000 a month, or $480,000 a year.

—Erica Jorgensen, author of *Strategic Content Design* (Jorgensen 2023) and staff content designer at Chewy

Qualitative research

The great thing about qualitative research is that it lets you understand users' attitudes, motivations, and thoughts as they move through a digital experience you're testing. User interviews and usability studies are the most commonly used forms of qualitative data collection. Chapter 8 covered user interviews. This section covers usability studies.

There are a lot of ways to measure the success of content, everything from qualitative to quantitative. If your organization isn't already doing it, I think it's a great idea to run A/B tests on content strings in isolation of other changes. When you do that, you really start to see the business impact words can have on the experience.

I also think there are cool interviews and surveys you can do around the impressions people have of the terms you used, and the names you've given things. That type of interview can help shed light on what's going on in your users' heads, giving you a lot more insight than an A/B test ever will. For more on this, I highly recommend *Strategic Content Design* by Erica Jorgensen.[1]

—Michael J. Metts, Principal Content Designer, Expedia Group Design System, and co-author of *Writing is Designing* (Metts & Welfle 2020)

Usability studies

These studies, typically conducted by a UX researcher, walk users through tasks, often performed in front of you. The person conducting the study has a guide created with your team that outlines the tasks and points of exploration. If you can, sit in on these studies. You'll learn a lot about how users move through your website as well as how they interpret your content. If you're not able to attend, make sure the sessions are recorded and that you receive a final report, which you can use to come up with next steps for creating the most effective content for users.

Another way to get involved is by participating in creating the script for the testing, also known as the moderator guide. This enables you to deepen the exploration of content into places where you think it will yield important results.

---

[1] *Strategic Content Design* (Jorgensen 2023)

One line of questioning asks users to read a piece of content and have them tell you what it means. For instance, if you have text on a button, you may ask them where they believe selecting the button will take them. The same holds true for links. For longer bits of text, such as error messages, help text, or blocks of content, the same exercise can yield invaluable results.

The final readout will contain findings from the research that include your content questions answered. Regardless of what tests you use, you'll gain insight into your work and where you can improve it. While planning which research to use, remember, there is such a thing as doing *too much* research, as Erika Hall says below.

For a particular question or study, reaching qualitative saturation means that doing additional interviews or observation isn't introducing any new themes.

In quantitative studies, statistical significance is a formula telling you the probability that your findings aren't just a matter of chance. It's important to note that just because results are statistically significant doesn't mean they explain anything real or true or useful.

Most people in their daily lives have a rough sense for how much information they need to make a major decision given the time frame and risks of being wrong. The design research process is just a way to formalize this and make it collaborative.

—Erika Hall, Author of *Just Enough Research* (Hall 2024)

TIDAL BANK

Case Study: User study

The team wants to share their work with users to make sure they've solved the problems they saw users having with the original Tidal Bank website.

The team works with Millie to set up an unmoderated user study where users interact with a clickable prototype of the new site to complete a few tasks and then answer a survey about the experience.

They use UserTesting.com to set up the test and ask users to learn more about the financial products Tidal Bank offers, select a type of account to open, and then open an account. After the user completes the tasks, they ask users to complete a survey rating the following statements from "Strongly disagree" to "Strongly agree." The survey includes statements such as:

- I think I would use Tidal Bank frequently
- I found the Tidal Bank website unnecessarily complex
- I thought the website was easy to use
- I was able to complete tasks quickly
- I felt confident using the website
- I imagine that most people would be able to use the website quickly
- I need to learn a lot of things before I can start using the website

They conduct the test with 15 participants and find that 77% were able to complete all of the tasks, and 90% of the participants said they strongly agree with the statements, "I thought the website was easy to use," and "I felt confident using the website." The team is happy and decides to move forward.

CHAPTER 12

Stakeholder buy-in

Presenting to stakeholders for buy-in or feedback means that you're showing the results of your work, so far, to get either their approval or feedback for your next iteration.

Before I get into presenting, I want to help you work with stakeholders more efficiently and collaboratively. Comrades in arms is what you're after, instead of arguments, egos, and right-ism. First of all, there is no "right" in content design. User experience—regardless of how much research, how many studies, and what the metrics show—can never be exactly "right." Here, I define right as right for users—a frictionless experience that is as easy as moving through air. That's a high bar, but a frictionless experience for every user is the holy grail.

Back to the stakeholders. First, let's look at who this cast of characters might include:

- UX and UI
- Development team
- Product managers and owners
- Accessibility and inclusivity partners
- Marketing
- Legal

That's a broad range of expertise to have at your disposal, but problems sometimes arise when stakeholders from different disciplines collide on their subject-matter facts or point-of-view. Many projects stall here or become incredibly inefficient, requiring review after review, frustration among peers, and sometimes angry words—none of which needs to happen.

How can you possibly understand the point-of-view of someone whose position in the company is completely different from yours? First, start by asking yourself these questions:

- What are their goals and responsibilities?
- How are you helping them achieve those goals and responsibilities, or how are you hindering?
- What do they know that you don't know?
- What do you know that they don't know?
- What are you looking for in terms of feedback?

If you don't know the answers to these questions, now is the time to set up a conversation with your stakeholders to get answers and build a rapport. This conversation will let your colleagues know that you care about their point-of-view and what they need in order to do their job effectively. If you have to set up meetings with each stakeholder, that's great. Go for it! The more relationship building you do, the better your project is going to go, the better your career is going to go, and, for a bonus, the more you will be doing to advocate for content-first design.

The conversation

When you truly listen to another person while they're talking, you're taking in more than their words. You're registering their mood as they talk, their body language. By remaining silent, you absorb a lot more information than you may have expected.

After you listen without interrupting, repeat back what you heard the other person say to make sure that you got everything. "What I heard you say was…." It's as simple as that. This makes your colleague feel that you're really trying to hear their concerns. This is relationship building in action, which requires empathy.

Empathy happens when you listen to your stakeholder and take in the information they're delivering and the way it makes them feel, which you can tell from their tone, their mood, and their body language as they talk. Once the stakeholder tells you that you've heard everything correctly, you can begin with your questions and concerns.

Now that you know where they're coming from, your conversation can take this into consideration. It might be to answer their fears, their bad experiences working with content in the past, their confusion over the process, or their goals and what they need in order to be successful. They might be determined to get their way, and by acknowledging their point-of-view, your empathy can potentially swing the tide in your direction or find ways to compromise while still achieving a seamless, user-centered content design.

This groundwork means that when you provide alternate solutions this person may be more likely to see the value in your point-of-view and that, together, you're more likely to arrive at a positive outcome. Your relationship can become collaborative rather than adversarial.

Capture your findings

After each interview, you may want to create a document in which you list what you learned so that you can keep it as a reminder or go back to your colleague for any missing information. These documents will help when you're working with other stakeholders now and in the future. They may also help other content folks as they kick-off their projects.

I think it's so useful to spend time understanding the incentives and motivations of other roles that you interact with and re-framing your job to show how you can help them. For years, I had a hard time working with product marketing managers. Sometimes we'd butt heads about in-app messaging, or they'd name a feature without considering how it works inside the product.

Finally, I realized that if I just openly acknowledged the ways we could help each other, it set us off on the right track. In a one-on-one with a new PMM I was to work with, I said, "You've done so much hard work to develop the messaging strategy to help users understand the value of this product. I'm here to help you find ways to deliver that message so it doesn't get lost in the mix when users open the app up."

I think that if you spend time understanding your partners and stakeholders, that helps you understand your own role and process more deeply, too.

—Andy Welfle, Principal Content Designer, Microsoft, and co-author of *Writing is Designing* (Metts & Welfle 2020)

TIDAL BANK

Case Study: The team

The team includes several product managers as well as a business partner and a marketing assistant in addition to the UX team. From the UX side, the team includes a UX lead, several UX designers and a UX researcher in addition to Anne, the content designer.

Each member of the team brings different skills and has different priorities. They want to identify those skills and priorities at the beginning of the project to ensure they understand any important goals or milestones early and that everyone can participate based on the skills they bring to the project.

For example, Kerry is a Sr. Product Manager who cares about achieving specific goals, including increasing conversion rates for visitors to the Tidal Bank website. She is skilled at gathering and understanding metrics so she could help gather initial metrics on how users are interacting with Tidal Bank and then compare metrics after changes have been implemented to understand how successful the project was.

They create Table 12.1 to help them understand the stakeholders and can go to the right people for the right feedback. It will also help them understand where each person is coming from, what they care about, and what they know about. And it's a great reference for names, roles, and email addresses.

Table 12.1 – Project stakeholders

Name	Role	Cares about	Knows about	Email
Anne Johnson	Business partner	Increasing conversion rates for Tidal bank	Communicating with leadership	Anne.Johnson@SS.com
Arez Abboud	Marketing Assistant	Setting up ads and marketing assets, ensuring brand consistency	Marketing copy, graphic design, brand consistency, ad tech	Arez.Abboud@tidalbank.com
Kerry Spinner	Sr. Product Manager	ROI, improving metrics, achieving goals, solving for problem statement, being heard	metrics, research, product details, customers, goals	Kerry.Spinner@tidalbank.com
Vamshi Nishar	Product Manager	Improved metrics, getting UX to solve the problem correctly, improved metrics, project success	metrics, research, history of product and its design, customers, project goals	Vamshi.Nishar@tidalbank.com
Roberta Rodrigues	UX lead	Connecting business needs and user needs, managing design projects		Roberta.Rodrigues@tidalbank.com
Santos Salas	UX Designer	User flows and user stories to design for, working well with stakeholders, achieving project goals	Design system, user habits in stories from previous work, Figma, and building a design rationale to present to stakeholders	Santos.Salas@tidalbank.com

Name	Role	Cares about	Knows about	Email
Amy Smith	UX Designer	Clear design guidance	User interface design, graphic design, following design system guidance. Working collaboratively	Amy.Smith@tidalbank.com
Millie Wu	UX Researcher	User testing, competitive research	User testing, competitive research	Millie.Wu@tidalbank.com
Anne	Content Designer	Being clear for the user both with language and flows, continuous testing, project success	UX Writing, design, accessibility, brand voice and tone guidelines, how to talk to these users	anne@tidalbank.com

How do you listen, empathize, and understand? Well, building a table like Table 12.1 is a great start but also check out Appendix B, *Tips on writing for a digital product* (p. 165), and you'll learn how to build deeper relationships with your peers.

Presenting

The more you understand what each person brings to the table, the more you can extract the knowledge you need from them to create the best content design. While everyone's contribution matters, it's up to you to sift through the ideas and lead the conversation so you can form a content design with a solid rationale.

Content-first design rationale

A solid content-first design rationale can help stakeholders understand why you have the content you're showing, and also let them know about the process by which you arrived at the content.

A solid rationale includes:

- The problem statement
- The strategic content decisions you made
- Why you made those decisions
- How you got to a final solution
- How your content design solves for the problem statement and helps achieve business goals

Your content-design rationale will provide stakeholders with a deeper understanding of the work and decision-making process so the conversation in the room becomes about strategy, avoids subjective opinions, and gets into thoughtful and strategic considerations. Before you present your rationale, you need to understand your stakeholders.

I aim to come in with an equally strong point of view as an open mind. I've found that succinctly communicating rationale for the decisions I've made while demonstrating a willingness to flex to new information or others' needs is the sweet spot for making myself heard and gaining trust as a respectful and respected partner.

—Selene De La Cruz, Sr. Manager of Content design at Robinhood

Understand your stakeholders

Everyone has their own agenda, and everyone has a piece of the puzzle you're trying to put together. The struggle to create alignment between business, accessibility, legal, marketing, and design can drive you crazy if you don't go into the presentation thinking of it as a collaboration with you as the facilitator and expert.

For each stakeholder, ask yourself:

- How are you helping them achieve their goals and responsibilities?
- What can you help them solve and what can't you help them solve?
- What do they know that you don't know?
- What do they need to see/hear from you in order to understand the content design solution?
- What are you looking for in terms of feedback from them?

Understanding your team helps you facilitate the conversation from a solid viewpoint with a strategy and research-based content design. Understanding what you need from each person helps you keep the conversation on track. Be sure to validate each person's comments by listening closely before you respond. Create an atmosphere of congeniality and curiosity by setting the example of not talking over people. This will help you get the most out of the meeting as you can.

For example, some stakeholders will not be swayed by content-research results, no matter how convincing they are to others on the team. When I asked Erika Hall if we should use research findings to sway stakeholders or to substantiate content designs, she had other ideas for you to consider. Her thoughts below illustrate that there are many ways to approach a problem.

Nearly everyone gets this backwards. Data doesn't change minds. And, in fact, if you introduce research results that strongly contradict the stakeholders' pre-existing beliefs, you may encounter what is known in psychology as the *backfire effect*. Not only will they reject the new information, they will hold even more strongly to their pre-existing beliefs. Many, many designers and design researchers have run into this phenomenon—showing up with a highly polished research report only to have it dismissed or ignored.

You need to get stakeholder buy-in and participation in the process before doing any research. **You do this by aligning on the goals and questions early and often.** If the stakeholder believes answering the research question is important to achieving their goals, and understands the proposed process to answer it, and agrees with the criteria for a sufficiently confident answer (never certainty, only confidence), then they will be interested in your findings. And they will be more likely to believe and remember your insights and recommendations.

Otherwise, you're in the position of answering a question they didn't ask and risk making them feel like you think you know better than they do. No one likes that feeling.

—Erika Hall, Author of *Just Enough Research* (Hall 2024)

Schedule the meeting

When scheduling the meeting, be sure to include an agenda. This not only helps you think through what you want to get across and what you hope to learn, but it also tells stakeholders what to expect so they can do any necessary up-front work beforehand to prepare.

A simple agenda can be a bulleted list that might look like this:

- Introductions
- Purpose of the meeting
- Expected outcomes
- The content design rationale and solution
- Discussion
- Next steps

You can also time-box each item in order to keep the meeting on track and make sure that you cover everything. Depending on your presentation, your time-boxing may look like this:

- Introductions – 5 min
- Purpose of the meeting – 5 min
- Expected outcomes – 5 min
- The content design rationale and solution – 20 min
- Discussion – 20 min
- Next steps – 5 min

The agenda should list all of the team members who will be in the meeting and contain a summary of your desired outcomes.

In addition to an agenda, assign someone from the team (or yourself) to take notes to distribute after the meeting for people to review. Sometimes, the quietest person in the room may review the notes and have a valuable insight to share with you outside the context of the meeting. Be clear, thoughtful, and thorough in your note taking.

Your notes will help you keep the team on track and provide a document to look back at if miscommunications arise. Notes provide a paper trail of decisions, plans, and changes so that you don't have to conduct the meeting again to solve the same issues.

Additional benefits of taking notes include:

- They reduce the amount of information you have to hold in your head so you can focus on the topic at hand.
- Some issues will be tabled or taken offline, and your notes can help you and your team respond thoughtfully based on what they discussed in the meeting.
- When you write down something another person said, it creates trust that you're taking them into consideration. That sense of recognition and validation helps others in the room feel valued as well.

While taking notes can be a pain in the neck, it's worth it. You can make it easier on yourself and those who will read your notes after the meeting by creating a clear outline, which you can organize by topic or by person speaking. Good notes help your team create a better content design, stay on track, and follow your lead.

Facilitate

As the meeting facilitator, you are in charge of keeping to the schedule and keeping the meeting on track. I've talked about the importance of listening to build trust and create a collegial environment. You'll need to use your listening skills to be an effective facilitator.

As the facilitator, it's important to keep in mind a few things:

- **Build relationships:** reach out to members of the team who have impact or influence on your project, get to know their roles
- **Empathize:** understand their point-of-view, respond to their concerns thoughtfully, and repeat their needs and requirements to make sure you understand and have consensus
- **Build trust:** show up, listen to others' visions, and base your rationale on metrics, research, and constraints
- **Deliver exceptional work:** make sure the information you present is the best you can create.

I find there are two key aspects to presenting work to stakeholders, both of which come down to trying to improve the relationship between you and the stakeholder rather than convincing the stakeholder to go one way or another.

The first is a deep understanding of what is keeping stakeholders up at night. What are their real needs? And where does your work intersect with those needs? Make sure that is the frame you put around your presentation. "Here is how accessible design is going to get you that bonus" versus "You should make this website more accessible on top of all your other work"

Second is presenting your work in a way that allows stakeholders to come to their own conclusions rather than pushing them towards your conclusion. This could be playing audio of a frustrated customer or leading a workshop around a problem you believe your design is going to solve.

Both of these approaches highlight finding common ground and focusing on solving problems rather than having a battle of wills over who's smarter. However you approach presenting your work, I encourage you to focus on building and enhancing the connection between you and the stakeholder rather than regulating that stakeholder's emotions and opinions; in the long run, the former is what really matters.

—David Dylan Thomas, Founder and CEO at David Dylan Thomas, LLC, author of *Design for Cognitive Bias* (Thomas 2020)

Letting colleagues know what kind of feedback you're looking for and what areas to focus on will help generate productive conversation.

Get the right feedback
As the facilitator, you direct the meeting. Here are some tips for getting good feedback.

- Ask participants to give everyone a chance to speak.
- Offer them the option of providing feedback offline by email or in a separate, one-on-one meeting. This gives them time to reflect on your presentation.
- Ask them to hold questions until you finish the presentation. This will help avoid interruptions and ensure that they get the whole picture.
- Let them know what's in and out of scope for discussion.

Present your solution
Present your content design using a rock-solid content rationale. A content rationale articulates:

- The problem you're trying to solve
- The context (where on the user journey the problem occurs)
- A summary of the results of the discover and define phases that informed your solution
- The strategic decisions you made
- Why you made those decisions
- How you reached your recommended solution

Lead with these points in your presentation. Basing your solution on facts will raise the discussion from one of opinions to one of strategies, which will help you reach a solution sooner.

Next steps
At the end of the meeting, get consensus on what the next steps are for refining content, working with UX and other colleagues, and sending content to development.

Send a follow-up email
A follow-up email will help the team stay on track by summarizing the feedback and findings from the meeting as well as next steps.

Unpopular opinion: sometimes, it is what it is.

I think there's a misconception that good outcomes mean changing people's minds every time. To influence. To advocate. To fight. To win?

And, sure, influencing skills are important, but I've learned one of the biggest skills is having the maturity to pick your battles, disagree and commit, let things go, learn, and do better next time.

Because if you're still dwelling on something that happened years ago but have no steps in place to improve things (or even improve how they impact you), the outcome will be the same.

The hard truth is: sometimes, no matter how many ways you explain and how right you think you are, people won't get it. Because people are complex human beings with their own frames of reference and focuses.

You can boil that up, keep banging the drum that you were right (you probably were). Or you can channel that fuel into working out what you learned from it. What opportunities can you take forward? How can you improve it? Can't improve it? That happens. Then, how can you make the best of the situation?

Sure, fight the good fight but try to fight it together. A battle of egos serves no one, and not every battle can be won. You'll nurture change faster than you'll dictate it.

Let's turn losses into learnings and recognize that we don't have control over every scenario, but it's success enough to have done our very best.

—Candi Williams, Content Design Director at Bumble, and Author, from a LinkedIn post

TIDAL BANK

Case Study: Present to stakeholders

Once the design team has the new Tidal Bank pages updated, they schedule a time to share their work with the important stakeholders and the larger Tidal Bank team. Because the stakeholders are not designers, it is important to outline the entire process and the rationale behind the design decisions the team made.

The team works together to create a presentation that clearly and quickly outlines their process. The presentation starts with stating the problems with the current Tidal Bank website that the team is trying to solve. Then they outline the design team's focus and their approach for updating the site.

They discuss the research that the team conducted and what they learned from the research. Anne discusses how updating the information architecture and hierarchy of the pages will help guide users through the website and make it easier for users to find the information they are looking for, as well as make it simpler to open a new bank account.

Then the team shows stakeholders the new designs and walks the whole group through why they made the choices they did.

After the design team finishes sharing the design, they open up a discussion with the stakeholders to get feedback and to clarify any questions or concerns stakeholders have.

If there are issues with the design that arise during this meeting, the team makes a plan for how to approach updates and adjust timelines.

Presenting: Roberta, UX lead; Anne, content designer; Santos, designer; Millie, UX researcher.

Stakeholder audience: Vamshi, product manager; Kerry, Sr. product manager; Felicity, business partner; Arez, marketing lead.

Roberta, the UX lead, opens the presentation.

—

A new design for Tidal Bank

Roberta: Welcome everyone! We are excited to show you the work we've done on creating a new design for the Tidal Bank website. We would like to go through the whole presentation first, so please hold your feedback and questions until the end.

Before we go into our new designs, we want to briefly take some time to review the current Tidal Bank website and outline some of the issues we encountered in our research and are addressing in our new designs.

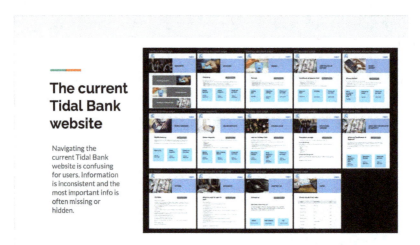

The current Tidal Bank website

Navigating the current Tidal Bank website is confusing for users. Information is inconsistent and the most important info is often missing or hidden.

Roberta: Here you can see some of the main pages of the current Tidal Bank website. The design feels outdated and there is a lot of missing information. Based on the metrics that our Analytics team gathered, users tend to drop off the website before they complete their tasks.

On the current Tidal Bank website, content is inconsistent and next best actions for the user are unclear.

Roberta: There are also clear inconsistencies in the language we use across the website, and we don't outline the steps users need to take to complete a task.

Our design goals

- Take a mobile-first design approach
- Consider the information architecture and content as part of the design process
- Create a new onboarding and new customer flow
- Inform users more clearly about products Tidal Bank offers to build knowledge and understanding
- Help users choose the correct product for their individual needs

Roberta: To address these issues, our design team outlined the following goals:

- Take a mobile-first design approach because most of our users come to the site via a mobile device
- Think through the information architecture of the site to create clear paths and an information flow that makes sense for our users
- Think through a couple of key flows, including the onboarding process for opening a new account
- Offer ways for users to build knowledge about the products Tidal Bank offers
- Make sure that our users can make informed decisions and find what they are looking for

As we go through the research and design process on the next few slides, Santos, Anne, and Millie will walk us through our findings and design decisions.

Our approach

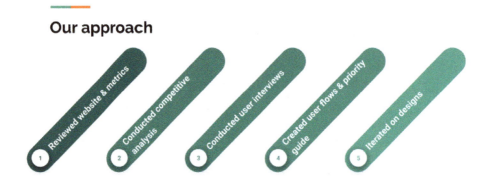

Santos: Here you can see the approach we took to tackle these goals. Importantly, we spent a significant amount of time doing research—including metrics gathering, competitive analysis, user interviews, empathy mapping, and content priority guides—before we started designing. This gave us a strong foundation for our design decisions and helped us ensure we were designing for the right problems and meeting our users' needs.

Key findings

53.3%

The bounce rate the Accounts page, a page considered high value

62%

Of users interviewed said they felt unsure of how to find what they were looking for.

1,500

The number of visits over the course of a month for the "What account is right for you?" page, ranking it very low.

Millie: Through our research we discovered a number of important findings. Here are several of our key findings. Importantly, one of the biggest themes of our research was that users too often don't find what they need, end up on pages they do not find useful, and, therefore, leave the site entirely.

Changes to the website structure

- Simplified structure
- Removed unuseful pages
- Reorganized home page
- Created new Opening Account page
- Created new quiz feature

Anne: To address the issues we uncovered in our research we decided to make several key structural changes to the website. Because users were getting confused or lost on our website, we decided to simplify the structure of the site, removing pages that weren't useful.

We also restructured the navigation of the website to make it easier for users to find what they are looking for. And we reorganized the home page to direct users more quickly to the items they say they are most interested in accessing.

To address issues with users not being able to complete the task of opening an account and completing the conversion process, we completely overhauled the "Opening an account page" to make sure users have everything they need before starting the process. This will help prevent users from leaving the flow without completing the account opening flow.

And finally, we created a new quiz feature to help users identify the right products for them. This will help inform users and promote different products based on user needs. We expect this to substantially increase conversion rates.

New design

We focused on restructuring some of the main pages to ensure users could find the information they were looking for quickly.

Santos: Here you can see the restructuring we did on the main pages to help make the pages more clear and easier to navigate.

Also, note that we designed mobile-first to make sure that our designs are useful and accessible on the screens that the majority of our target audience uses to access our website.

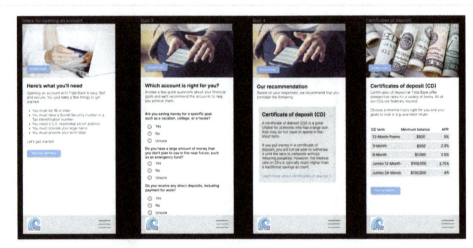

New pages include a page explaining what's needed to open an account and a quiz to help users identify the right account for them.

Anne: Here we've used content design to enrich our "Opening an account" page. By outlining the requirements for opening an account up front, we can keep users from dropping out of the process early.

You can also see an example of the quiz page. Clear, conversational questions help inform users and can also help them find the right products.

We created a new template for account pages to maintain consistency and to simplify pages.

Santos: Notice the new format of our product pages. We created a clear template to use for all of our product pages to keep things consistent and to keep information in formats that are easy for users to digest.

Because we have a template, we can also create clear guidelines for using this template—both for the design and the content—to ensure new product pages follow the same format in the future.

Key content design principles

- Clear content hierarchy
- Simple Calls-to-Action (CTAs)
- Chunked information
- Simple, explanatory language
- Consistent tone and voice

Anne: Finally, we want to point out a few of the key content design principles that we followed in this redesign process:

- Restructured the website and eliminated duplicate and irrelevant pages in order to create a clear content hierarchy.
- Updated buttons and other calls-to-action with clear and simple language.
- Broke up the content on the product pages into chunks to make it easier for users to read and digest.
- Simplified language where possible to ensure all of our users could understand the ins and outs of our products.
- Updated language so that we use a consistent voice throughout the website based on our style and brand guidelines.

Feedback?

—

Roberta: Thank you to the entire design team for their work on this project. We have confidence that these new designs will be much more user-friendly and will help Tidal Bank achieve its business goals.

Now, does anyone have any questions or feedback they'd like to discuss?

The group begins a feedback and general discussion.

Felicity: We will need to watch the metrics once this is released to make sure it meets all of our KPIs for lowering bounce rates and increasing completion rates. I am concerned that there are still going to be a lot of calls from users who have questions. What if this doesn't work? Can we go back to the old design?

Roberta: That's a great point, and we will keep an eye on those metrics once the design is launched. If we see that the metrics after a few weeks aren't meeting our KPIs, we can review together and see if there are ways for us to improve the design. We don't plan to go back to the old design. Instead we will work together to improve the new design.

Vamshi: We'll need to check in with the development team about upkeep on that quiz. It's a great idea, but we need to make sure the tech behind it will work and can be maintained easily.

Anne: That's a great call out! Let's have a meeting with the development team to discuss the technical capabilities and how we can build this quiz in a way that's sustainable.

Kerry: I'm concerned that the simplified language doesn't adequately describe the products. What if there are compliance or legal issues with the language?

Anne: Let's check with the legal team to review the language and discuss some key guidelines for the template so that if we add new product pages in the future, we can ensure they meet the legal team's requirements.

Arez: What if we want to promote specific financial products? Is there space for that on the website?

Santos: Yes! Because the home screen has a more modular design now, we can easily add promotional sections to the site. Let's talk about what potential requirements you have for a module like that and then we can build a template.

Roberta: Thank you all for your participation today! We will meet again in a few weeks to discuss how things are going and to see if we need to update the design based on the metrics we have agreed to gather.

CHAPTER 13

Advocate for content-first design

Just get started and the advocacy surfaces naturally. Reach out to your customers however you can—online interviews, in-person interviews, surveys, diary studies—and you'll inevitably discover surprising, even gob-smacking insights about which words work and which don't, and that will help you immediately improve and clarify the content on your site or app.

That improved clarity and engagement leads to more successful, happier customers and a less overwhelmed customer-service team, because they'll need to field fewer emails or chats from confused or unhappy customers.

The best part is being able to measure improved revenue and cost savings for your business from clear, more engaging content. A stronger-performing website or app means a more successful business. This is exciting and empowering! The irrefutable business impact of great content speaks for itself and will naturally create support for more time and effort put toward content research in your organization.

—Erica Jorgensen, author of *Strategic Content Design* (Jorgensen 2023) and staff content designer at Chewy

Advocating for content-first design within your organization requires a strategic approach, open communication, and regularly demonstrating the value it brings.

In my experience, advocating for content-first design requires educating stakeholders and decision makers about the benefits to the company of designing from a content-first approach. This can be accomplished through the user research, user interviews, and measuring the efficacy of your content using the techniques covered in this book.

When a team I was managing wanted their stakeholders—product, business, UX design, accessibility, legal, and marketing—to understand why it was essential to allocate time and money for content-first design, they created a presentation that showed how content-first design works, what information it yields, and how they use that information to address the issues outlined in the problem statement.

Most of the people attending the presentation had no idea about the role that content could play in informing product and design. At the end of the presentation, and after asking clarifying questions, the team was convinced that content-first design would yield measurable improvement in product engagement.

Including stakeholders in content definition tasks, such as empathy mapping and user interviews, helps educate them about the process and lets them see firsthand how the process leads to different results than if a designer sat down and developed the user experience in a content vacuum. The more you include others, the more you earn a seat at the table and the more willing they become to take on a new process for design.

Build relationships with stakeholders

Get to know your stakeholders and what their responsibilities, goals, and challenges are. Learn how they think of content and figure out ways to educate and inform them without forcing your agenda. In spite of the challenges, some stakeholders understand the importance of bringing content design in early. In that case, you're a lucky content designer!

Get a seat at the table where decisions are made by bringing your stakeholders into the content-first design process with:

- Invitations to presentations and workshops
- Roadshows educating that content-first design is more than copywriting
- Requests for their input

Wherever you can demonstrate the value of content-first design, spread the word far and wide. Use metrics and usability studies that demonstrate the value of putting content first. And, introduce your team to UX stakeholders so that they can get this important relationship started.

Demonstrate the benefits of content-first design

Being able to demonstrate the benefits of content-first design can make a big difference with leadership and other project partners. This starts with a plan for measuring content-first design against business and user goals. Measurable results demonstrate the efficacy of putting content-first, which will help your team and your stakeholders shift toward a new, unfamiliar process.

You may need to educate a lot of people about the content-first process and its benefits before you can get agreement to implement it at an organizational level. That's why it's important to keep chipping away. Invite decision makers and stakeholders to your presentations on the process and include them in your workshops, user research studies, and user interviews. Bring them into the process so they see, firsthand, the benefits users are experiencing.

Start small. Present to your UX team and stakeholders to get feedback, then branch out and present to other teams until you're ready to present to the broader UX team. Include higher-level decision makers such as content directors and UX directors. Understand their points-of-view and take those points-of-view into consideration. Their first concern will be about allocating money and extra time. That's why you need to demonstrate measurable results.

If your content team is still fighting to prove its worth, you can begin by demonstrating the measurable results of your work through metrics, usability studies, and user interviews. Remember, the proof is in the pudding. Show your value to stakeholders and decision makers, and they will begin to follow.

Keep fighting the good fight. Learn how other content teams have gotten on the company radar and keep doing your work to get there. Find encouragement in the success of others. Talk to content folks at other companies or people in the community who you admire and learn, learn, learn. Persistence is the key to any kind of success. Go for it!

Be prepared to address concerns or objections that can arise among UX peers and stakeholders in general. These concerns can include resource constraints, time limitations, or skepticism about the impact of the process. Anticipate these objections and provide concrete examples and evidence to counter them.

Work with leadership

Engage with leadership and secure their support for content-first design. Clearly articulate the strategic importance of content in achieving organizational goals and demonstrate how putting content-first can contribute to those objectives by creating the most meaningful conversations with users. Having leadership backing your efforts can significantly influence the adoption of content-first design practices throughout the UX organization.

One way to get leadership on board is to celebrate even the smallest wins. These could include an effective redesign of a vital page, great user feedback, improved metrics, or an overall uptick in usability studies. Leaders love to have simple, effective stories that demonstrate a win.

Advocating for content-first design in your organization requires persistence, effective communication, and a commitment to demonstrating tangible benefits. By aligning content-first design with organizational goals, educating stakeholders, and showcasing measurable success, you can champion the cause and contribute to creating more user-centric and effective digital experiences.

CHAPTER 14

Implement, learn, keep going

Now that you've learned one process[1] for content-first design, what it can do, and how it can improve the design process, you're ready to decide if this is right for you and your organization.

Getting a seat at the table

There can be challenges to introducing content-first design to your organization, especially if you don't have a seat at the table. One way to get that seat is by generating behind-the-scenes successes that are measurable and that demonstrate the efficacy of your work.

For example, find a pain point in a flow that isn't performing well and measure and test the content. Look at metrics, then get a UX designer and developer on board to implement simple content changes to improve the experience. Then measure until you get positive results that you can share.

Another idea is to get to know the team through content-first exercises such as workshops or presentations—something that lets them know you understand their point-of-view, yet educates them about the power of content-first design and the benefits of bringing you in earlier.

It might be an inch-by-inch process, but stick with it! You'll get there! The most important thing is to make the process and tasks as engaging and non-threatening as possible to stakeholders. Seek to understand their point-of-view and encourage others to do the same when you set up the ground rules for your content-first process.

A phased approach

A phased approach is one way to start. Rather than pulling out the double-diamond, or whatever design paradigm you choose to work with, begin by introducing a few techniques for content-first design. For instance, you could start with an inventory, since that's a useful exercise even if you don't continue on to a full content-first project.

---

[1] Remember, the process may vary depending on your team's particular needs, content, business culture, and preferences, but the basic concepts are consistent.

The benefits of starting with a content inventory include the following:

- You'll know what you have
- You'll be set up for an audit
- You can navigate the user flow and evaluate what pages are needed and what are not

You can create workshops, as demonstrated in the case study for Tidal Bank. You have a wide range of tools, techniques, and processes to choose from for workshops, including:

- Content inventories and audits
- Empathy mapping
- Priority guides
- User research
- Designing and implementing
- Testing

Be creative

This book lays out *one* vision of how to implement a content-first design process. As you start incorporating these processes into your work you may find that some techniques work better for you than others. Also, you can adapt these ideas to work better for you.

Even if you don't implement a content-first design process, I hope you find the tools in this book useful. Whatever you do with the knowledge you've gained, even just thinking about content first can create a better user experience for everyone.

Community

My main goal in writing this book is to get the conversation going about content-first design. What can we learn from each other? How's it working at your organization? There's so much to talk about here, let's get that conversation started. You can reach me at info@contentfirst.com.

I've started a blog at contentfirstdesign.com/blog/. This is a vehicle to discuss methods we're using to move the practice forward, whether it's with ways that you're improving content-first design, implementing content design systems, or whatever you've come up with or discovered and are digging into. It's a place to ask questions, share ideas, and share accomplishments and challenges. I hope you'll join the blog and continue your work moving content forward.

TIDAL BANK

Case Study: Project outcomes

Once the developers implement the design and everyone has had a chance to review the work, the development team launches the new design to users.

Through the process of redesigning the Tidal Bank website, the team learned a lot. Through the content inventory and content audit, they gained a better understanding of how some pages on the site worked better than others and why. Through the competitive analysis, they gained insight into what Tidal Bank's competitors are doing and how they might incorporate what others are doing. During the user interviews they learned how real users interact with their website.

As they built out priority guides, they learned about organizing the content on a page and how information architecture can shape the design. Through the design and prototyping process they learned how to bring their ideas to life and how to iterate and improve as they go.

They were able to distill their work into clear ideas and articulate why they made each design decision. And finally, testing their new designs with users helped them validate their work.

The Tidal Bank design team's work is done...for now. But as users start to use the new Tidal Bank website, the data analytics team will watch to see how the new site performs. If they notice something that's not working, they will come back to the design team to see if there are ways to address these new user problems, and the cycle begins again!

The art of the difficult conversation

A difficult conversation drove me into the field of mindful leadership. I was in a stakeholder meeting with a product partner, a business partner, a UX designer, and a visual designer. We were delivering a redesigned page that included visual elements as well as text content.

The product partner was agitated on the call, so I immediately didn't like him. When the page review came around to content, this guy threw me under the bus, demanding that the content be done sooner and saying that I was to blame for holding up the project. He said this in front of twelve colleagues.

I left the meeting angrier than I'd ever been over a work call. I couldn't believe how pissed off I was. At the same time, I thought "now everyone knows I'm a loser, an impostor" and "I'm no good for this job," then "that guy's a jerk, he doesn't even know me." Back and forth I went.

The inevitable flame-a-gram came next. I sat down at my computer to begin composing, determined to send my scathing response to the entire list of meeting attendees and thereby embarrass the hell out of this miscreant, but even in that moment, I stepped back. Flame-a-grams have a way of bouncing back and biting you where it hurts. I stormed around and raged in my brain until work was over. After a long walk, I simmered down.

On my walk, I realized that my anger was only hurting me, so in addition to having a negative outcome to the meeting, I had to suffer with feelings of anger, insecurity, and fear.

I spoke with my friend Walter, and it didn't take long for me to relay the conversation in great gory detail, emphasizing all the ways in which this person was truly a jerk. Walter listened patiently, nodded his head, and pointed me directly to the Search Inside Yourself Leadership Institute's book, *Search Inside Yourself* (Tan 2014).

That night I went online and ordered the book. After reading about SBNRR (Stop, Breathe, Notice, Reflect, and Respond), I reflected on that awful conversation. I did the first thing right: **Stop**. I stopped myself from writing a flame-a-gram to the team. Now I could **Breathe** and **Notice** how the conversation made me feel physically, emotionally, and mentally. Just thinking about the conversation made my heart race, my face flush, and my emotions run all over the place along with my mind.

When I **Reflected** on why this was happening, and why my reaction was so extreme, I realized that I was insecure about my skills and experience. While I had built this wall of confidence—some of it real, some of it not—around me as protection, I carried an insecurity with me. When I dug deeper, I realized that it wasn't only about work, but about myself as a person, as if I didn't deserve to be here, in this house, with this job, living the good things in my life.

When I realized all of this, I began to think about the person who'd offended me and thereby triggered me so deeply. What was driving him, what did he need in that moment? I wondered if his boss was on him to deliver faster, if he felt insecure somehow and needed to come across as in control, or if he needed to take down someone in order to feel okay about himself or impress others. The more I dug into my perceived enemy's point-of-view, the less angry I felt. Could he have handled himself differently? Yes, but he had his own challenges within himself that he was not conscious of and that had nothing to do with me.

So how did I **Respond**? I chose not to write the flame-a-gram and didn't bring it up with this person individually. While I sat back, it turned out that other people responded to him, and he was taken off the project. The reason I did nothing was because he didn't seem like the kind of person who would be open to the kind of conversation I had been trained to have.

Being mindful of ourselves can improve our success at work, at home, and any place where we interact with other people. Plain and simple.

How to approach a difficult conversation

Difficult conversations are an art, and once you learn how to conduct one, you can help create positive outcomes from even the most volatile, disheartening situations. Of course, there are those who, no matter what, are simply not available or present enough to conduct a difficult conversation, but more often than not, coworkers and clients want to resolve hard issues.

In the course of our careers, we will run into difficult conversations where we are on the receiving end of another person's uncomfortable communication style, and we may often leave these conversations feeling angry, resentful, hurt, sad, or scared. You name it, the negative emotions start pouring in. I want to offer an alternate approach to difficult conversations, one that includes active, compassionate listening, and thoughtful feedback, as well as a conversation in which you reach resolution and both parties leave feeling happy, successful, optimistic, maybe even proud.

Sounds pretty good, doesn't it? Let's get started.

But…What does this have to do with content-first design?

Having the skills to navigate difficult conversations allows us to interface with other people to create positive, productive outcomes. Compassion for others drives much of what we do in content-first design. We try to understand our users and have empathy for them. What better way is there to do that than to first understand ourselves and where we're coming from so that we can understand our users and where they're coming from.

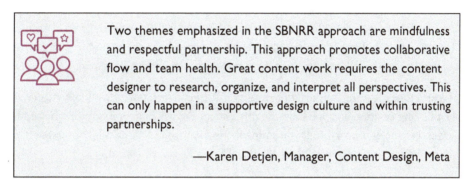

Two themes emphasized in the SBNRR approach are mindfulness and respectful partnership. This approach promotes collaborative flow and team health. Great content work requires the content designer to research, organize, and interpret all perspectives. This can only happen in a supportive design culture and within trusting partnerships.

—Karen Detjen, Manager, Content Design, Meta

Before we dig in, let's take a moment to settle ourselves.

Three deep breaths

Take three deep breaths and focus on the feeling of your breath as it moves into your nostrils.

The minute it takes to do this simple exercise will pay off by improving your focus and attention, relaxing your mind and helping you feel more present and grounded.

Let's do it again—three deep breaths. Let your focus settle on the feeling of your breath moving into your nostrils.

How do you feel now? Do you feel more relaxed? Focused? Or is it something else? Notice for yourself how you feel in this moment, take stock without judging. You're not supposed to feel any one way. You may feel things in your body, your mind, your breathing. If not, that's okay, too. Taking a few breaths can help you manage difficulties of all varieties.

Try taking a few deep, conscious breaths in the following situations:

- Before an important conversation
- When you feel triggered
- Before you transition from being at work to being home
- When you have the urge to check your phone or social media

Notice how it makes you feel. Does the conversation go differently, do you feel more grounded in the conversation or better able to listen to the other person? If you're triggered by something, your feelings will be boiling, but rather than react to those feelings, take a few deep and conscious breaths to settle yourself down and consider how to proceed in a positive way.

Why is conscious breathing important before a difficult conversation? If you know you're going to have this conversation and you have the time, breathing will help you be more present to yourself and to the other person. If you suddenly find yourself in a difficult conversation, don't react immediately. Take a breath. Slow down and listen before you speak.

Identify a practice example

Think of a difficult conversation that you've had or that you're about to have and imagine the positive outcome you want(ed) to achieve. Use this as a learning example throughout these pages.

Difficult conversations can become a battleground instead of a space for innovation. Content that works with difficult conversations frames the conversation as "this vs. that" instead of "product and user vs. the challenging topic" and focuses on a solution instead of the problem. It's essential for content designers to learn how to have those difficult conversations in analog before trying to create difficult navigation in a digital space.

—Kelcea Barnes, Freelance Content Designer, former Senior Content Designer Upwork, 2022 Meta Design Fellow

Prepare for the conversation

If you know you're about to have a difficult conversation, taking the time to prepare yourself can mean the difference between a negative outcome or a positive outcome. With a little forethought, you can enter a difficult conversation in a more understanding mindset. If you don't know that the conversation is coming, and you're side-swiped and unable to achieve the positive outcome you want, you can turn this negative outcome around using the steps I'll cover here.

First and foremost, check in with yourself. How are you feeling? Are you agitated, nervous, angry, or feeling defeated? Is your mind racing with all the ways in which you're "right" and winning is the outcome you want. Or are you calm—your mind at ease. If the former is true, be sure to take some time out to breathe. With some deep breaths to calm your mind, you'll realize that there's no reason to get upset. You'll handle the conversation thoughtfully, using these methods.

Once you feel more at ease, consider who you'll be having the conversation with.

- What do they want?
- How do they feel?
- What's at stake for them?

While you may not know the answer to all of these questions, simply asking yourself the questions will help you get yourself into a curious mindset toward the other person. Any successful outcome requires two willing participants to listen and learn from each other.

Stop. Breathe. Notice. Reflect. Respond. (SBNRR)

I worked on this method for having challenging or difficult conversations with the Search Inside Yourself Leadership Institute, a mindfulness department at Google, after participating in their original version of the workshop. More and more companies are setting up such departments in order to achieve more positive business outcomes.

For example, LinkedIn has a strong focus on mindfulness ethics and teaching. Their CEO, Jeff Weiner, talks a lot about compassionate leadership, mindfulness, and the importance of conscious breathing in order to help create a smoothly running organization.

SBNRR is a method for helping you respond effectively in a difficult conversation. Consider how it can help the situation you've been keeping in mind as an example.

- **Stop:** When you notice yourself becoming triggered, just stop. Don't react for one moment.
- **Breathe:** Take a deep breath. Focus on your breath to calm your mind and body.
- **Notice:** Notice what's happening in your body, your emotions, and your thoughts. Are these feelings static or are they changing from minute to minute or from second to second.
- **Reflect:** Why is this happening for you? Why are you triggered? Is there a history behind your feelings? Is there a self-perceived inadequacy involved? Without judging it, bring this perspective into the situation.
- **Respond:** How can you respond to create a positive outcome? You don't have to actually do it, just imagine the most kind, positive response. What would that look like?

Take a moment now to think about the difficult conversation you are recalling or anticipating. How could this method have helped you? What do you learn from reflecting this way? Once you get used to practicing this method, it can become reflex, but it takes time and attention. Don't expect to get it right in every conversation you have or even once—just keep practicing.

The 5 steps

These steps will help you navigate even the most difficult conversation, whether at work or at home. Read through them and practice them when you have a quiet moment. These five steps are taken from a workshop with Google's Search Inside Yourself Leadership Institute. You may want to check out their book by Chade-Meng Tan, *Search Inside Yourself* (Tan 2014).

Step 1

Imagine a conversation you're about to have or that you've had in the past

- What happened that you need to address or that you addressed in the past?
- What emotions do you or did you have?
- What stories did you tell yourself about what this meant about you?
- How did this confront your identity, or how you think/thought about yourself?
  - ► Am I competent?
  - ► Am I a good, worthy person?

Step 2

Check your intention and decide whether to raise the issue

What do you hope to accomplish by having this conversation? Is it a productive intention (trying to solve the problem) or is it non-productive (proving a point or blaming the other person)?

- Sometimes not raising the issue is the way to go
- If you do raise the issue, try shifting to a mindset of curiosity, learning, and problem solving

Step 3

Start from the objective "third story"

This is the way things happened from the perspective of an objective third-party who is aware of the whole situation. Looking at things from this perspective can help you find common ground with the person you want to talk with.

Step 4

Explore their story and yours

- Listen
- Empathize and actively practice an attitude of seeing similarities and offering kindness
- Share your story
- Explore together how each of you perceives the same situation differently
- Reframe the story from one of blame and accusation to one of learning about how each of you contributes to the situation and emotions involved

Step 5

Problem-solve

- Create solutions that meet each other's concerns and interests
- Find ways to continue keeping communications open and taking care of each other's interests

Even if, at first, you explore only one step, you're off to a good start. I run through these steps when I've had or am about to have a difficult conversation. This exercise can shift you from feeling fear, anger, and anxiety to being calm, centered, and open.

When to use the 5 steps

When you're anticipating a difficult conversation that you want to prepare for or when you're thinking back over a past conversation and want to learn from it, mentally reviewing these steps can help you create positive outcomes. The more success you have using these steps, the more likely you will be to continue practicing them.

Listening exercise

This exercise will help you put the 5 steps to work. Find a partner—anyone in your proximity who is willing to try something new. Using the tools you've learned through SBNRR, have a conversation that goes like this:

- You talk while your counterpart listens
- Your counterpart says "what I heard you say is…," and you give feedback until you're satisfied that they've heard everything.
- Switch roles and do the exercise again.
- Now have a free-flowing talk about how the exercise went, what each of you experienced, and what you felt.

Here are a few suggestions for topics to discuss:

- Talk about a time when you overcame a challenge
- Talk about someone in your life you particularly appreciate and why
- Talk about anything that feels meaningful to you

You can also try this approach in the moment when you feel comfortable and you've tried it a few times. Remember…

This process takes practice!

It gets easier as you begin to use it, so hang in there. Breathe…

Tips on writing for a digital product

I've learned a lot about UX writing over the years just by practicing the trade and seeing what worked based on research and testing. Writing for UX is a vast field and much has been written on the topic. Certainly much more than I provide here. I encourage you to look at other books to learn the ins and outs, but here I present an overview to help you with the tasks you've learned about in this book. Maybe it will inspire you to take a course in content design or UX writing if you're new to the field.

Here's a quote from my favorite UX book out there:

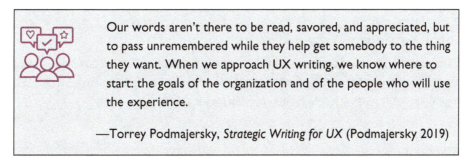

> Our words aren't there to be read, savored, and appreciated, but to pass unremembered while they help get somebody to the thing they want. When we approach UX writing, we know where to start: the goals of the organization and of the people who will use the experience.
>
> —Torrey Podmajersky, *Strategic Writing for UX* (Podmajersky 2019)

No matter where you are in your content design career, be sure to read Torrey's book. Regardless of your experience level, you'll benefit from her wisdom and strategic approach to content design.

Another favorite is *Content Design: Research, plan and deliver the content your audience wants and needs*, by Sarah Winters and Rachel Edwards.[1] The book is written using the principles of content design throughout the text, so it teaches you as you read.

Getting started writing for digital products

When you are first tasked with writing a piece of content, it's important to understand where the user is coming from and where they are going. If you're dealing with a link or a button, what happens immediately before and after they click on it?

---

[1] *Content Design* (Winters & Edwards 2024)

Understanding the user's journey through your website is key to content-first design. How can you plan for a journey if you don't know where you're going? It seems simple, but this understanding of from/to is often overlooked. When you understand the conversation you're trying to have with the user, you can work with designers to create meaningful experiences.

The elements of content-first writing

User-centered

The content you create should always be focused on the user first, based on discovery and research, which I cover in Chapter 3, *The discover phase* (p. 25). Understanding and empathizing with the audience for your work can lead to insights such as pain points, task goals, motivations, feelings, and frustrations.

Simple and clear

Plain language will ensure that your content is clear and readable. The US government site plainlanguage.gov provides a ton of information for writing in a digital environment and lists a set of guidelines for writing in plain language. These include:

- Write for your audience
- Organize the information
- Choose your words carefully
- Be concise
- Keep it conversational
- Design for reading
- Follow web standards
- Test your assumptions

Using plain language, avoiding jargon, and keeping sentences and paragraphs short will all contribute to the readability of your work.

Visit plainlanguage.gov for more information on how to effectively write in plain language for your users.

Readable (readable.com) and Hemingway (hemingwayapp.com) offer tools to measure the readability of your text based on grade-level—which will depend on your audience—as well as your brand, tone, and voice guidelines. Hemingway offers additional benefits as well, including grammar and spell checking.

Scannable

Organize your information to be scannable. This helps users find what they're looking for faster. Use headings, sub-headings, bullet points, and short paragraphs that give users the essentials. According to nngroup.com, scannable content can produce a range of benefits, including:

- Faster completion times for users
- Reduced comprehension errors
- Reduced recollection errors
- Better understanding of the site's structure
- Fewer users leaving the site
- Higher likelihood of return visits
- Improved search engine optimization (SEO)

Visual hierarchy

A visual hierarchy is directly related to scannability. The most important information comes first, the next most important comes second, and so on down the line. When you create scannable content, make sure that the most important information for the user comes first on the page. Even at the sentence level, put the punch at the beginning. People skim larger pieces of text or scan down the left side of a paragraph, jumping over sentences within a paragraph to find what they need. Use that knowledge to create effective content that users will want to engage with.

Figure B.1 – Visual hierarchy[6]

---

[6] https://www.linkedin.com/posts/jacksonyew_what-do-you-see-first-activity-7197621052183650304-cV46/
© 2024 Jackson Yew. Used with his permission.

Figure B.1 is from Jackson Yew's LinkedIn page (https://www.linkedin.com/in/jacksonyew/), which, along with his website (jacksonyew.com), has some great information on content design. Jackson designed this particular image to illustrate how a user's eye traverses a page.

This is a good point to stop and develop your own opinion about whether you agree or disagree with the statements in Figure B.1. You may agree completely. Or not. Either is fine. Personally, I agree with his assessment of how users scan or read a page.

Here's what Jakob Nielsen, of the Nielsen Norman Group, has to say about how readers consume information online:

People rarely read Web pages word by word; instead, they scan the page, picking out individual words and sentences.

In research on how people read websites,[8] we found that 79 percent of our test users always scanned any new page they came across; only 16 percent read word-by-word.

As a result, Web pages have to employ **scannable** text, using

- highlighted **keywords** (hypertext links serve as one form of highlighting; typeface variations and color are others)
- meaningful **sub-headings** (not "clever" ones)
- bulleted **lists**
- **one idea** per paragraph (users will skip over any additional ideas if they are not caught by the first few words in the paragraph)
- the inverted pyramid style,[9] starting with the conclusion
- **half the word count** (or less) than conventional writing

—Jakob Nielsen, "How Users Read on the Web" (Nielsen 1997)

---

[8] "Applying Writing Guidelines to Web Pages" (Morkes 1998)
[9] "Inverted Pyramids in Cyberspace" (Nielsen 1996)

Visual hierarchy critique

Figure B.2 is a screenshot from mrbottles.com. You can see from what we just covered that people are not likely to read the huge paragraph. And that's just the beginning! How might you make this page more scannable based on your instincts, experience, and possibly the illustration in Figure B.1? Give it a try and see what you come up with. Consider what the site is trying to get across or provide. Then, break those ideas down using the concepts we've covered so far.

What might you use from the double-diamond approach to improve the site?

What are your recommendations? You can send me email at sarah@contentfirstdesign.com. I'd love to hear them.

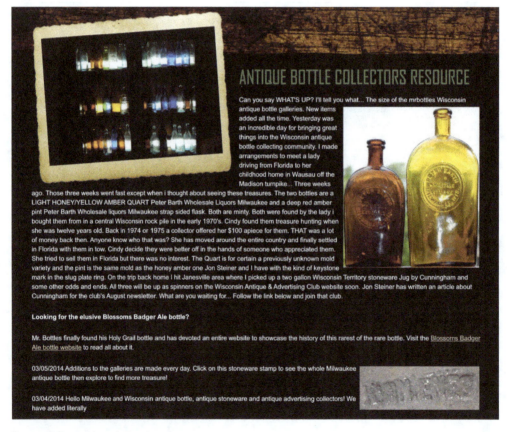

Figure B.2 – mrbottles website

> *****Spoiler alert*****
>
> My approach to the mrbottles site would be to read the text and edit it down. I would create a headline that highlighted what the site was about. Then provide meaningful headings and sub-headings to guide the user through the flow of inform-ation. Anything I could do to make the content more succinct and better guide the user would help. For example, I might add a menu and a description of the re-sources they offer. In addition, lists can help create a more scannable site.

Relevant

Is the content relevant to the users' understanding of information or the task at hand? If not, you don't need it. Evaluating relevance can help you create more concise content that serves the user's needs and that users want to engage with.

Consistent

Consistency is critical to users as they work their way through your content. Use the same tone and voice throughout. Also use consistent vocabulary. If you call something a dog in one place and a canine in another, users become confused. As they navigate a page, users parse information and create mental models. Once you've used the word dog, readers will expect the word dog, and anything else will slow them down or make them stop.

>
>
> **Mental model**
>
> A mental model is a model the user creates in their mind about your website, encapsulating what they expect to see based on what they've seen so far on your site and on other sites.

Consistent formatting helps users create a mental model so they become accustomed to finding things in the same place every time. A good example of consistency is an e-commerce site like target.com. Every page that has the same purpose uses the same format. For example, product details pages all have the same information. Check it out sometime on any site you come across and see how well they adhere to this important guideline.

Accessible

...experiences are only usable when they're accessible.[12]

Accessible content is understandable, readable, and findable for anyone regardless of their abilities. It's a way for organizations to reach a wide audience and take down barriers to communication for everyone. For example, blind or visually impaired people often use screen readers, which expect certain elements and styles to be in place in order to parse a page. By ensuring that content is accessible, you begin to create inclusivity and equity. You also need to think about people who may not be like you in all kinds of ways, for example education, culture, language, or race.

Some examples of accessibility requirements for content:

- Text alternatives for images
- Headings to create structure
- Bullet and numbers to create scannability
- Headings and summaries for tables
- Clear, descriptive links

For example:

- Not descriptive: To schedule an appointment, **click here**
- Descriptive: **Schedule an appointment**

 Web Content Accessibility Guidelines (WCAG) are globally recognized recommendations for making digital content more accessible to users with disabilities. You can find them here:

https://www.w3.org/TR/WCAG22/

---

[12] Candi Williams, Content Design Director at Bumble

"As someone who has led and worked on accessibility projects for years, I could write a book on the myths around it.

But let me just say this: accessibility isn't hard or expensive.

What's expensive is: continually choosing to ship inaccessible MVPs [minimum viable products] and having to go back over five years of these when lawyers come knocking with 8-figure fines.

What's hard is: not being able to access basic user experiences because someone couldn't spare 10 minutes to create accessible text or test with assistive technology.

It is beyond misguided to suggest AI will solve accessibility. But if people leaned into accessibility with even 10% of as much enthusiasm as GenAI, experiences would be infinitely better for millions of people. And companies' profits would grow millions.

Do I long for a day where I can make 25 silly little avatars of my face in the style of an 80s yearbook? No. But I do long for the day people would acknowledge the deep bias behind separating usability and accessibility and accept that experiences are only usable when they're accessible."

—Candi Williams, Content Design Director at Bumble[13]

---

[13] https://www.linkedin.com/posts/activity-7170853902999896064-SAot

Inclusive

 "Inclusive content is a commitment to equity. It means providing a safer experience for more people. It's not a badge of honor you earn with a single initiative or intention—it's an ongoing and intentional effort. Like accessibility, inclusivity is essential to creating meaningful user experiences that reach everyone."

—Rease Rios[14]

Differences come in many forms including the following:

- Sight
- Hearing
- Physical abilities
- Gender
- Race
- Age
- Sexuality
- Ethnicity
- Neurodiversity

Writing for equity and inclusion means using plain language, simple sentence structure, and formatting that's easy to read and understand. Avoid writing that contains gender connotations, such as mankind, manpower, or mothering. Try humankind, employees, and nurturing instead.

 Here are two guides for making your writing more inclusive: "Avoid Gender Bias in Writing" (Western Michigan University 2023) is a writing style guide for avoiding gender bias, and "A Guide to Writing Inclusive Language and Copy" (Payne 2021) looks more generally at inclusivity in writing.

---

[14] "How to create inclusive content that welcomes a wider audience" (Rios 2021)

Neurodivergent[15] users may have trouble focusing, processing a complicated sentence or idea, following a linear process, and understanding abstract language or thinking. How do you write so that neurodivergent users aren't left out?

First of all, use plain language that is broken down into small sentences and small ideas. Complex ideas must be broken down carefully so that they can be understood by all users. This helps reduce cognitive load and makes it easier for everyone, not only those who are neurodivergent, to make sense of what you're saying.

In addition, clear headings and sub-headings that lead all users through an experience seamlessly are a must. As I've covered, breaking down pages into digestible pieces benefits everyone. Audio or visual aids can help, and a glossary if you absolutely need to use jargon or technical terms.[16]

Writing inclusive content and creating inclusive experiences requires inclusive hiring practices to build teams that are diverse in terms of race, gender identity, age, cultural identity, and more. There's so much research showing how diverse teams create more successful products. If a team feels comfortable talking through the hard topics like how a product could exclude or cause harm, they'll be much better equipped to address those topics.

I'll also say it requires support from management at higher levels. People have to be willing to make something inclusive even if they can't immediately see short term business benefits or boosts to the metrics they've been tasked with. It requires a backbone—being willing to do the right thing.

—Michael J. Metts, Principal Content Designer, Expedia Group Design System, and
co-author of *Writing is Designing* (Metts & Welfle 2020)

---

[15] Neurodivergent is a non-medical term that refers to the diversity of cognitive functioning in people. It encompasses both strengths and challenges. For more on neurodiversity, see this article: "What is Neurodiversity?" (Wiginton 2023)

[16] "Making Content Accessible to a Neurodivergent Audience" (Reed 2023)

Writing microcopy

Microcopy consists of small pieces of content in the user interface that help guide users through the digital experience as smoothly as possible. It can come in many forms, including Calls to Action (CTAs), button text, empty states (which occurs when no content appears in the UI or on a whole page), placeholders, labels, instructional text and error messages.

When drafting microcopy, keep it concise and clear. The fewer words the better.

Let's look at an example. In Figure B.3, the text **Find out how** on the orange, oblong button is an example of microcopy.

Figure B.3 – Microcopy

Types of microcopy

Microcopy includes error messages, instructional text and tooltips, notifications and alerts, onboarding content, dialogs/modals, dashboards, empty states, forms, transactional emails, and success messages. The following sections look into each of these.

Error messages

Error messages need to be clear and actionable for users.

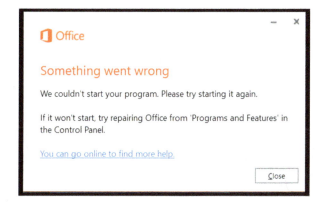

DO:

- Be clear about what went wrong, when possible. If the issue is a complicated technical situation, it may be better to be vague
- Give the user steps to fix the problem

DON'T:

- Dismiss or blame the user
- Unnecessarily explain the issue
- Be vague about the way to fix things

Instructional text & tooltips

Instructional text is text that appears next to something that may need additional explanation. A tooltip is instructional text that is hidden until a user clicks on or hovers over it.

DO:

- Add contextual guidance for things that aren't obvious
- Keep your tips short and sweet
- Make tooltips dismissable

DON'T:

- Hide essential information in a tooltip
- Make your tooltips longer than a sentence

Notifications & alerts

Notifications are messages that appear on a user's screen to let them know about something that has changed. Error messages and success messages are subsets of notifications.

DO:

- Be clear about the type of alert
- Be clear about next actions if needed
- Keep Calls to Action (CTAs) short and descriptive

DON'T:

- Overload the user with alerts

Onboarding

Onboarding is a guided walkthrough or tour of product features. This often appears the first time a user opens a mobile app or other digital tool. Onboarding messages can also appear when an application has been updated significantly or has added new features.

DO:

- Use onboarding flows to help users new to a product or screen
- Allow onboarding flows to be dismissable
- Use images and text to get to the point

DON'T:

- Overuse onboarding or show it every time a user logs in
- Make your onboarding too long

Dialogs/Modals

Dialogs and modals are screens that appear over the main digital experience to ask the user for more information, confirm a selection, or indicate a change.

DO:

- Use modals for important information
- Use modals to get your users' attention
- Make sure modals are dismissable

DON'T:

- Overuse pop-ups and modals

Dashboards

Dashboards are screens that share important data for the user. They often have graphs, progress bars or other graphical representations of information. A good dashboard helps users understand complex information at a glance.

DO:

- Use dashboards to show a lot of information in an easy to view manner
- Make dashboards customizable
- Make sure legends and labels are easy to read

DON'T:

- Use a dashboard unnecessarily

Empty states

Empty states are screens that do not currently have any of the expected content in them. For example, if you go to your email inbox but you have no messages, the inbox would show an empty state. This can be an important place to set user expectations for what they should see when the screen is not empty.

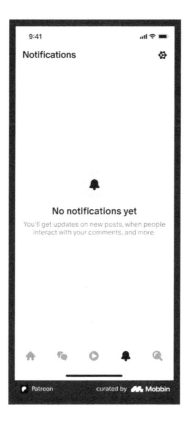

DO:

- Use empty states to set expectations
- Keep empty state copy short and to the point

DON'T:

- Add a lot of extra or unnecessary information to empty states

Forms

Forms are screens that ask users to enter bits of information into different fields. The fields can be open-ended boxes, drop-down menus, or selection boxes. It's important to make sure users understand what information is being requested of them and what format it should be in, because forms have a high rate of errors.

DO:

- Use clear, consistent names for form field labels
- Use helper text for form fields
- Use specialized form fields when possible; for example, a dropdown for dates
- Break up forms into chunks
- Use inline errors to help users fix information quickly

DON'T:

- Build overly long forms
- Be vague about why each field is being requested
- Ask for unnecessary information

Transactional emails

Transactional emails are sent to a user to convey information about a transaction that the user has initiated with a company. For example, if the user has purchased a product online, they would receive a transactional email telling them that the order has gone through, another when the item has shipped, and another when the item has been delivered. It is important that the transactional information is front and center in these types of emails.

DO:

- Include all order information
- Be clear about how a user can track an order
- Include pricing and quantity information
- Give users contact information

DON'T:

- Add superfluous information to the email
- Make it difficult for users to track their order

Success messages

Success messages are messages that appear on a screen when the action a user has taken has been successful. These can be great moments for celebration and delight. A good success message can reinforce a user's experience.

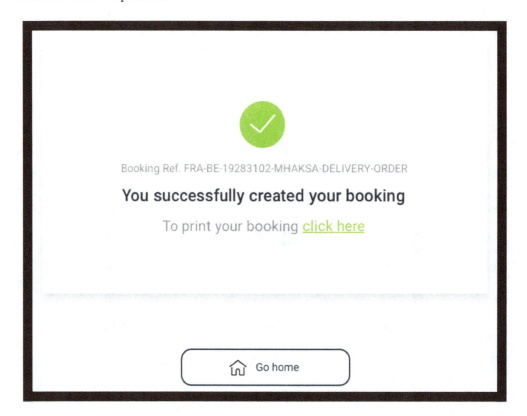

DO:

- Confirm that users have successfully completed a task
- Keep success and confirmation messages short and simple
- Use Calls to Action (CTAs) sparingly

DON'T:

- Allow confirmation messages to disappear without the user dismissing them

Why is microcopy important?

Using only a few words, microcopy keeps users moving and interacting with your product or site. Small but mighty, this content can do some heavy lifting for your product by, for example, letting users know what they can (and can't) do next, explaining a new feature or function, or providing reassurance when faced with an error.

Since context is everything when it comes to microcopy. Ask yourself the following questions before you start writing:

- Who are the users?
- What do they want?
- What motivates the users?
- What concerns or reservations do they have?
- What are we offering users?

You need to know the answers to these questions and understand the voice and tone you're writing in before you begin writing microcopy. These are questions you should be able to answer about any piece of content, but because microcopy has so few words in which to convey a message, you need to really bear down and get to the bottom of answering those questions.

Microcopy is critical to the success of your project, and you should pay attention to microcopy related metrics both before delivery and after.

Microcopy isn't just a bunch of standalone words that hop onto the page. Everything goes into consideration—where the user came from, where the user might go next, and where the user might have problems along the way.

If you don't understand how your content is organized and laid out throughout the whole experience, you'll never get those few words of microcopy to work. Remember, microcopy lives as a part of the whole experience, not a slice of it!

—Laura Lifshitz, CVS Health

Dig deeper

The Bibliography (p. 189) contains an extensive list of articles and books related to content-first design. Here are some references that focus on the material in this chapter, plus pointers to several writing-related workshops:

Books

- Torrey Podmajersky, *Strategic Writing for UX* (Podmajersky 2019)
- Kinneret Yifrah, *Microcopy* (Yifrah 2019)
- Michael J. Metts and Andy Welfle, *Writing is Designing* (Metts & Welfle 2020)
- Janice (Ginny) Redish, *Letting Go of the Words* (Redish 2012)

Workshops

- Contentfirstdesign.com
- UX Content Collective, uxcontent.com
- UX Writing Hub, uxwritinghub.com

Bibliography

Bailie, Rahel Anne, and Noz Urbina. 2012. *Content Strategy: Connecting the Dots Between Business, Brand, and Benefits*. Denver, CO: XML Press.

British Design Council. "The Double Diamond." https://www.designcouncil.org.uk/our-resources/-the-double-diamond/

Brown, Tim, and Barry Katz. 2019. *Change by Design: How Design Thinking Transforms Organizations and Inspires Innovation*. Revised and updated edition. Harper Business.

Clemens, Drew. 2012. "Design Process In The Responsive Age." https://www.smashingmagazine.com/2012/-05/design-process-responsive-age/

Dunn, Beth. 2021. *Cultivating Content Design*. A Book Apart. https://www.bethdunn.com/book

Gibbons, Sarah. 2018. "Empathy Mapping: The First Step in Design Thinking." https://www.nngroup.com/-articles/empathy-mapping/

Grigoryan, Sophie. 2024. "Usability vs User Experience: What is The Difference?" https://userpilot.com/-blog/usability-vs-user-experience/

Hall, Erika. 2024. *Just Enough Research*. 2024 edition. Mule Books.

IDEO. "What's the difference between human-centered design and design thinking?" https://-designthinking.ideo.com/faq/-whats-the-difference-between-human-centered-design-and-design-thinking

Jorgensen, Erica. 2023. *Strategic Content Design: Tools and Research Techniques for Better UX*. Rosenfeld Media.

Land, Paula Ladenburg. 2023. *Content Audits and Inventories: A Handbook for Content Analysis*. 2nd ed. Denver, CO: XML Press.

Metts, Michael J., and Andy Welfle. 2020. *Writing is Designing: Words and the User Experience*. Rosenfeld Media.

Morkes, John, and Jakob Nielsen. 1998. "Applying Writing Guidelines to Web Pages." https://-www.nngroup.com/articles/applying-writing-guidelines-web-pages/

Nielsen, Jakob. 1997. "How Users Read on the Web." https://www.nngroup.com/articles/-how-users-read-on-the-web/

Nielsen, Jakob. 1996. "Inverted Pyramids in Cyberspace." https://www.nngroup.com/articles/-inverted-pyramids-in-cyberspace/

Payne, Scarlett. 2021. "A Guide to Writing Inclusive Language and Copy." https://boldist.co/usability/-writing-inclusive-language/

Podmajersky, Torrey. 2019. *Strategic Writing for UX: Drive Engagement, Conversion, and Retention with Every Word.* 1st ed. O'Reilly Media.

Redish, Janice (Ginny). 2012. *Letting Go of the Words: Writing Web Content that Works.* 2nd ed. Morgan Kaufmann.

Reed, Stacy. 2023. "Making Content Accessible to a Neurodivergent Audience: A Guide for Technical Writers." https://storiesfromtheherd.com/-making-content-accessible-to-a-neurodivergent-audience-a-guide-for-technical-writers-1965668372e

Rios, Rease. 2021. "How to create inclusive content that welcomes a wider audience." https://webflow.com/-blog/inclusive-content

Sanford, Biz. 2017. "Words and the design process: Greetings from a friendly content strategist." https://-medium.com/shopify-ux/words-and-the-design-process-f41472a249fb

Tan, Chade-Meng. 2014. *Search Inside Yourself: The Unexpected Path to Achieving Success, Happiness (and World Peace).* Reprint edition. HarperOne.

Thomas, David Dylan. 2020. *Design for Cognitive Bias.* A Book Apart.

Vinney, Cynthia. 2023. "What is human-centered design? Everything you need to know." https://-www.uxdesigninstitute.com/blog/what-is-human-centered-design/

Western Michigan University. 2023. "Avoid Gender Bias in Writing." https://wmich.edu/writing/genderbias

Wiginton, Keri. 2023. "What is Neurodiversity?" https://www.webmd.com/add-adhd/features/-what-is-neurodiversity

Winters, Sarah, and Rachel Edwards. 2024. *Content Design: Research, plan and deliver the content your audience wants and needs.* Lulu.com.

Yifrah, Kinneret. 2019. *Microcopy: The Complete Guide.* 2nd ed. Nemala.

Index

Colophon

About Sarah Johnson

Sarah Johnson, a content design leader and teacher with over 20 years of experience, has worked for industry leaders such as Fidelity Investments, Banks of America, TIAA, CVS, and Bentley University User Experience Design Center.

Sarah is the author of six books, including *Content First Design*, and the founder and director of ContentFirstDesign.com. Content-first Design, the company, offers content services built on actionable, data-driven insights, and workshops designed to enhance practical skills in areas such as content design, AI integration, and more.

About XML Press

XML Press specializes in publications for technical communicators, content strategists, marketing communicators, and managers. We focus on concise, practical publications concerning content strategy, management, and XML technologies. You can reach XML Press at:

This book and other XML Press publications can be purchased directly from XML Press or through online retailers worldwide. Discounts are available for bulk orders.

Internet: https://xmlpress.net
Email: publisher@xmlpress.net
Phone: (970) 231-3624